MERGERS & ACQUISITIONS

CYBERSECURITY

MERGERS & ACQUISITIONS CYBERSECURITY

THE FRAMEWORK FOR MAXIMIZING VALUE

LAWRENCE GRANT

Mergers & Acquisitions Cybersecurity
The Framework for Maximizing Value

Published through Thales Publishing

Book Formatting: Creative Publishing Book Design

ISBN: 978-0-9896668-8-6 (Paperback Edition)
ISBN: 978-0-9896668-9-3 (Hardcover Edition)
ISBN: 978-0-9896668-1-7 (eBook Edition)

To my loving and inspiring wife, Michele, thank you for your unwavering support, conversations, and encouragement through the many late nights and wee morning hours that made this book come to fruition.

Contents

Foreword

When I started my journey in law school, I was instinctively drawn to the idea of corporate law. At first, my primary motivator was the same as every young law student fresh from a liberal arts institution and saddled with debt: money. But as I learned more about business and the legal foundations which underpin it, I came to appreciate corporate law for what it truly is—a regime for protecting value. Value is more than just the dollars and cents that are exchanged in any given deal. It represents the time and energy that founders put into their company, the aspirations acquirers have for its future, and the benefits that employees, customers, and stakeholders reap when its value is realized.

It was this realization that thrust me into the world of Mergers & Acquisitions (M&A) and Capital Markets. I became invested in learning how to safeguard my clients' value. My time in law school and private practice has taught me techniques for how to achieve that goal. Due diligence, iron-clad contracts, and strict adherence to regulatory frameworks have long been considered essential for successful corporate transactions.

Given the sums at stake, it would be imprudent for buyers and sellers to not identify and manage threats before they manifest. Unfortunately, the battlefield is ever-changing. Even as these risks constantly evolve, it is my job to peek over the horizon and update the regime as needed. My experiences as a corporate lawyer alongside my conversations with business owners, fellow attorneys, and experts such as our esteemed author, have led me to conclude that poor cybersecurity represents one of the greatest classes of risks to corporate transactions in the 21st century.

Lone wolf hackers, sophisticated criminal networks, and even nation-state actors have run amuck in the business world. And nowhere is this pain felt more acutely than in an M&A deal. Billions of dollars have been lost due to cybersecurity compromises, and that number is only projected to grow.

And yet, it's hard to blame founders, firms, and financiers for their oversight. As a first-hand participant in the mania that often accompanies high-value M&A deals, I can say that they all have their hands full. There are so many pins to juggle in an M&A transaction that it should come as no surprise that this issue can form an Achilles' heel for even the best of businesses.

Stakeholders need actionable advice if they are to properly navigate the world of cybersecurity. They need practical steps on how to secure their systems, protect their data, and bring their transactions to a smooth close. Cybersecurity needs to go from being an afterthought to a comprehensive strategy on par with legal and accounting matters. Only an industry-wide shift will suffice.

This book represents the first big step in that shift.

This book is the perfect companion for any individual who is part of the M&A value chain. It will help cybersecurity experts in understanding the risks posed to deals of this type. It will help founders and owners increase the value of their business by minimizing risk. It will help private equity firms and other buyers accurately price targets by accounting for that risk. And it will help lawyers in drafting legal documentation to allocate that risk. It is a handy tool that gives a global overview of what's at stake, provides clear steps on how to address cybersecurity, and explains the value saved in doing so.

But more importantly, the MAC-F framework is the beginning of a new counter-offensive against the cyber intruders of the world. It is a sword and shield designed to stop bad-faith actors and empower stakeholders so that they can protect their value and ensure it reaches its maximum potential.

Nigel Frank, Esq.
Associate Attorney, Pryor Cashman LLP

Chapter 1

Introduction

This book is written for a diverse audience, including owners, private equity firms, family offices, Mergers and Acquisitions (M&A) professionals, corporate leaders, cybersecurity specialists, and anyone involved in generating economic value through buying, selling, or merging companies. It aims to protect M&A transaction values and protect stakeholders' reputations by proactively managing cyber risk and mitigating its damage before, during, and after the transaction. For the sake of conciseness throughout this text, the terms "transaction," "deal," or "M&A" may be interchangeably used to refer to an outright acquisition, company merger, or company sale, depending on the context.

In addition to the relevant explanations for the cybersecurity domain, this book contains real-world accounts of my professional journey that shed light on the subject. Given the sensitive nature of M&A transactions and the importance of confidentiality, generic names and identifiers are used to safeguard the interests of the entities involved.

Cybersecurity is a comprehensive discipline that protects information processing systems, digital infrastructure, networks, and data from theft, damage, and unauthorized breaches. It represents a confluence of technology, processes, and human resource strategies implemented to protect an organization's digital assets and business value.

If we ask different individuals to define cybersecurity, we may obtain as many opinions as there are respondents. Therefore, we will start with a simple definition universally agreed upon regarding the elements that constitute cybersecurity, formerly information security. The core principles of cybersecurity are reflected in the acronym CIA, commonly called the CIA Triad. Not to be confused with the other CIA, the United States Central Intelligence Agency.

Confidentiality

This principle, aligned with data privacy, ensures that information and information processing systems are accessible only to authorized entities. This is a defense against unauthorized access to or exposure to information.

Integrity

This principle protects against unauthorized changes or alteration of information, information processing systems, and infrastructure. This is a defense against unauthorized modifications or alterations.

Availability

This principle ensures the continuous functionality and accessibility of information processing systems, infrastructure, and associated networks by the business entity. This is a defense against system downtime and business interruptions.

For stakeholders, understanding the principles of the CIA triad and applying them is necessary for preserving the economic value of M&A transactions in an environment of globally persistent cyber threats.

Cybersecurity is a nonnegotiable aspect of enterprise risk management. During the past three decades, cybersecurity breaches have affected the reputations, valuations of stocks, revenues, and legal standing of numerous organizations. JP Morgan Chase's CEO, Jamie Dimon, has said that the cybersecurity threat is the "biggest threat" to the financial services industry, with his company spending nearly $600 million annually on security.

For organizations, the recognition that effective cybersecurity management is critical to business health has led to the establishment of dedicated roles, such as the Chief Information Security Officer (CISO) and departments with teams of cybersecurity professionals. These experts diligently work to protect the organization's digital assets, ensuring that operations run smoothly and data remain secure. Nevertheless, many small-to mid-market companies and even Fortune 10 companies experience staffing and budgetary challenges protecting against daily cyber threats.

Although day-to-day operations are essential, there is a realm where even the most experienced security experts may face unfamiliar territory: mergers and acquisitions. The M&A process presents unique challenges, intricacies, and factors that may not arise during typical business operations.

The cybersecurity team may be skilled in managing a company's security program, executing security initiatives aligned with IT and business strategies, and defending against cyber threats during business-as-usual operations.

However, safeguarding a company during the tumultuous process of M&A requires a different, tactical approach, whose goal is to ensure a successful exit or merger for the seller and inoculate the buyer from unforeseen risk, financial loss, and even deal collapse.

I do not suggest that the security approach during M&A should be an isolated endeavor. Rather, it should build on the in-place security program's functioning aspects and adjust its tactical focus to support the success of M&A transactions. In summary, timelines are short during M&A, the monetary value is at its peak, and the strategic outcomes are weighty.

Every transaction, whether a merger of two entities or an outright acquisition or sale, has intellectual property, data, infrastructure, and other digital assets supporting the deal's value. If these assets are not protected or integrated securely, they may be vulnerable to compromise. Cybersecurity threats can threaten the economic and strategic foundations upon which the parties justify the acquisition.

This book is categorically framed in this context. Recognizing the gap between traditional cybersecurity management and M&A demands, this book aims to equip stakeholders, either from the buy side or from the sell side, with insights, strategies, and a prioritized cybersecurity framework to prepare for the cybersecurity challenges of mergers and acquisitions with confidence.

A side-by-side examination of steady-state security management and cybersecurity challenges during M&As highlights key differences. The table that follows shows the differences between the two cybersecurity contexts.

Cybersecurity Impact	Business-As-Usual Profile	Mergers and Acquisitions Profile
Valuation	Standard Metrics: Business valuations are determined by tangible assets, market position, and revenue.	Multiplier Effect: M&A valuations are often a multiplier on the company's revenue, for instance, valuing the business at 10X its annual revenue. The revenue multiplier raises the impact of cybersecurity risks.
Time frames	Strategic: Security program implementations can take 2 to 5 years to achieve best-practice maturity levels.	Tactical: The entire deal process can take 6 to 12 months from initial contact to closing. The due diligence phase may be less than 90 days.
Environment	Steady-state: Operates within defined parameters with established systems, networks, and processes.	Fluid: Introduces unfamiliar elements, including new systems, different data handling practices, new stakeholders, and external entities.
Risk	Fixed scope: The reputation, assets, finances, and legal standing of a single entity.	Expanded scope: Reputations, assets, finances, and legal standing of two or more entities.
Focus	Internal: Primary focus on safeguarding internal assets.	Internal and External: Increased depth and frequency of external engagements, from sharing sensitive data to integration coordination.

The most salient difference between business-as-usual cybersecurity management and cybersecurity in M&A is the outsized impact that a security incident or lapse can have on a transaction's value and success. In the context of M&A, cybersecurity is not just about defense; it is an integral component that influences the overall success of the transaction. The following case study of Verizon's 2016 acquisition of Yahoo Inc. illuminates the impact of cybersecurity issues during the M&A process.

In 2016, Verizon announced its intention to acquire Yahoo's core Internet business for an estimated $4.83 billion. Once an Internet pioneer, Yahoo saw a decline in its market share and sought a strategic acquirer. For Verizon, the Yahoo acquisition was seen as a fit for its overall strategy to grow its digital marketing advertising business.

As the acquisition proceeded, a significant event disrupted it. Yahoo disclosed two massive data breaches that occurred in 2013 and 2014. These breaches have compromised the data of more than a billion users, making them among the most impacting cybersecurity incidents in history.

Deal Impact

After the data breach disclosure, Verizon renegotiated the terms of the deal. The final acquisition price was reduced by $350 million, bringing the total to $4.48 billion. This reduction was a direct consequence of the cybersecurity incidents, highlighting the enormous financial impact of such breaches on M&A valuation.

Legal Liabilities

Apart from the reduction in the purchase price, both companies agreed to share regulatory and legal liabilities arising from the data breaches. Yahoo would be responsible for 50% of any cash liability incurred from government investigations and third-party litigation unrelated to the Securities and Exchange Commission (SEC). The liability from shareholder lawsuits and SEC investigations would be entirely borne by Yahoo.

Reputational Damage

The breaches tarnished Yahoo's reputation and triggered skepticism regarding its user base's loyalty and overall brand strength. Reputational damage was another factor that influenced the reduced valuation.

Valuation Impact Analysis

Yahoo announced revenue of approximately $4.96 billion for the fiscal year ending December 31, 2015. Verizon's initial acquisition proposal was $4.83 billion, suggesting a valuation that was roughly equivalent to Yahoo's revenue.

After the revelation of cybersecurity incidents, Verizon renegotiated its offer, setting a purchase price of $4.48 billion. This adjustment represented an approximate 7.25 percent reduction from the original offer at a whopping $350 million.

Chapter 2

Cybersecurity Deal Threats

Shortly after becoming the Chief Information Security Officer of a medium-sized multinational firm, I was thrust into managing a crisis situation: an ongoing distributed denial of service (DDoS) attack that affected our primary service of hosting e-Commerce platforms for leading retailers. This incident was not an anomaly. The firm had been grappling with an uptick in these attacks, which consumed our technical resources and jeopardized our primary operations.

The timing was particularly sensitive. The company was in acquisition talks with a larger corporation. With each DDoS attack, our company's valuation in this deal was at stake.

Beyond the immediate technical implications of the DDoS attacks, the financial strain they placed on the firm would be untenable if the situation persisted without an imminent solution. Incident handling costs were mounting, including additional pay for staff working overtime to find a resolution and maintain uptime, penalties for breach of customer service agreements, and revenue lost due to platform downtime. There was an urgent need for a sustainable and cost-effective solution.

After examining various technical options and our contractual Service Level Agreements (SLAs) with customers, I settled on a novel approach. Collaborating with our legal team, we engaged our customers, provided transparency about the issue, and obtained agreements to modify their SLAs. Taking inspiration from the "force majeure" legal clauses, where cataclysmic unforeseeable circumstances, for example, earthquakes, prevent an entity from fulfilling a contract, we decided to treat DDoS attacks as "extraordinary events beyond the company's control" akin to "Acts of God."

My solution was that for each DDoS attack targeting a client's hosted site, our company would commit technical resources for 30 minutes. The affected site would transition to a specialized DDoS mitigation provider if the disruption continued beyond that period. DDoS mitigation providers allow entities to redirect bogus Internet traffic to the service provider, who maintains a purpose-built infrastructure that can absorb the enormous bandwidth resources the attack consumes. The linchpin of this strategy was that the customer would bear any additional costs arising from this extended mitigation. This approach would thwart the cyberattack, remove the associated expenses from our balance sheet, return revenues to normal levels, preserve our customer relationships, and prevent erosion of the acquisition's value.

The proposal for revised customer contract terms was met with unanimous approval from our board and executive team. An impressive 95% of our clients embraced the updated contracts, understanding the importance of ensuring the uninterrupted operation of their e-Commerce platforms. This strategy extended beyond cybersecurity acumen and was made possible through the collaboration of the legal team overseeing the transaction.

With the execution of this strategy, we strengthened our operational resilience against DDoS attacks and stopped the financial hemorrhage caused by platform downtime and SLA performance refunds for customers. It was also instrumental in preserving our company's valuation during the acquisition.

The digital transformation of businesses has brought unprecedented opportunities and inevitably, a myriad of cybersecurity threats. Consider the following M&A scenario between two entities and the impact that various threats can pose to the deal.

Never Age Biotech is a groundbreaking biotechnology firm that specializes in innovative treatments that promise to slow the aging process and inhibit fat accumulation. With annual revenues of $75 million, their proprietary treatment has garnered attention within the medical wellness community and has paved the way for them to be recognized as a prime acquisition target in the biotech industry. Their unique formulations can revolutionize the way people age and vastly improve the health of people with weight problems, making them a valuable asset in the biotech market.

High Returns Private Equity is a major investment entity known for its strategic acquisitions and for delivering impressive returns to its stakeholders. Their modus operandi is identifying companies with high growth potential, particularly those that bring innovation to their respective sectors. High Returns Private Equity aims to capitalize on emerging markets and innovative technological advancements in the medical field by acquiring firms such as Never Age Biotech. The offer price of High Returns Private Equity is $1.125 billion, reflecting a 15-fold revenue multiple for Never Age Biotech.

As these companies discuss the possibility of an acquisition, there is immense opportunity for value realization, but it is accompanied by challenges, including those related to cybersecurity. To provide insight, the following examples illustrate common threat scenarios and their impact on the deal's success. They are contextualized within the CIA triad of security principles discussed earlier.

Deal Threat: Distributed Denial of Service (DDoS)

Never Age Biotech systems are brought down by a massive hack. DDoS attacks overwhelm the company's online systems by flooding them with illegitimate traffic, often rendering them temporarily inoperative for lengthy periods.

Availability Compromise

This threat can halt all online business activities, affecting treatment orders, customer service, and other operations.

Deal Impact

Prolonged service interruptions might raise concerns for High Returns Private Equity regarding Never Age Biotech's preparedness against cyber threats. Such doubts can lead to renegotiating the acquisition terms or deal termination.

- If public trust is eroded, post-acquisition revenue predictions may require a downward adjustment. High Returns Private Equity may view the company as a riskier investment, influencing the deal's valuation.

- The news of the attack may be widely publicized, leading customers and investors to lose faith in the company's viability.
- These attacks may attract regulatory scrutiny, and the company may face higher compliance costs or fines.

In 2016, Dyn, a primary DNS provider, faced a massive DDoS attack that made major websites such as Twitter, Reddit, and CNN inaccessible.

Deal Threat: Intellectual Property Theft

A hacker breaks in and steals the company's formula. Intellectual property theft involves the unauthorized access, use, and dissemination of a company's proprietary information.

Confidentiality Compromise

Unauthorized access to the unique treatment formulas of Never Age Biotech can cost the company its market position, as cheaper knockoffs using the stolen formula are released.

Deal Impact

- The unique selling point and core value proposition of Never Age Biotech would be compromised, making the company less attractive to High Returns Private Equity or any other suitor. The acquisition price may be renegotiated downward if the deal survives.
- Legal complications can delay the acquisition, and the associated costs might again force a reevaluation of the deal's financial terms.

- Never Age Biotech might have to engage in legal battles to protect its intellectual property rights, incurring substantial costs.

$1 billion in value and 700 jobs were lost almost overnight when the Chinese company Sinovel paid a reported $2 million bribe to an Australian employee to steal the source code of American Superconductor's wind turbine in 2011.

Deal Threat: Advanced Persistent Threats (APTs)

State-sponsored hackers break into Never Age Biotech systems to manipulate data and the source code. APTs are long-term targeted cyberattacks in which intruders gain and maintain unauthorized access to a network, often undetected for extended periods.

Confidentiality, Integrity, and Availability Compromise

APT groups might steal valuable data over time and manipulate data and source code, thereby compromising current operations and future R&D projects.

Deal Impact

- Discovering such a breach, especially post-acquisition, can lead to large financial losses and a negative reputational impact. High Returns Private Equity might face unexpected costs and challenges, impacting their Return on Investment (ROI).
- This breach would gravely compromise Never Age Biotech's proprietary advantage, leading High Returns Private Equity to reconsider the acquisition's terms or feasibility.

In 2023, Clorox reported an attack by ATP hackers in public filings, a breach that disrupted its systems. The attack caused a decrease of approximately 25% in quarterly revenue for the company, translating into approximately $49 million in losses.

Deal Threat: Social Engineering

A lab employee is tricked into uploading the company's flagship formula to the site of a hacking group. Social engineering attacks manipulate people to divulge confidential information or perform actions that compromise security, often exploiting human emotions such as trust or urgency.

Confidentiality Compromise

Attackers employing social engineering techniques might deceive employees into revealing sensitive information about Never Age Biotech's proprietary treatment.

Deal Impact

If such an attack is successful, the proprietary secrets of Never Age Biotech may be exposed or misused, diminishing the company's unique value proposition. High Return Private Equity might renegotiate the deal's terms, seek indemnities, or even re-evaluate the acquisition's overall merit. Furthermore, the revelation of such an attack could undermine trust between the two entities, complicating the acquisition process.

In 2015, Ubiquiti Networks reported a loss of $46.7 million due to a social engineering attack. Cybercriminals impersonated executive staff in communications with the finance department, leading to unauthorized fund transfers.

Deal Threat: Insiders

A disgruntled employee leaks confidential information about High Return Private Equity's offer to purchase the company. Insider threats originate from individuals within an organization, such as employees or contractors, who might misuse their access maliciously or inadvertently. Insider threats can be lethal if confidential information about the deal is leaked to or by employees or executives.

Confidentiality Compromise

An insider with malicious intent or someone who has been compromised could leak confidential data, sell R&D results to a competitor, or carelessly leave their laptop unattended in a coffee shop and have it stolen.

Deal Impact

Insider threats can be a clear and present danger to an acquisition's success. If High Return Private Equity discovers internal vulnerabilities post-acquisition, it could mean unexpected costs and management challenges, affecting their expected Return on Investment (ROI). Furthermore, any internal compromise at Never Age Biotech could alter the relationship, leading to revaluation or collapse of the deal.

In 2016, the UK-based software firm Sage faced an insider-driven data breach that affected its customers. The incident demonstrated the risks posed by those with inside access and their ability to compromise customer data and relations.

Deal Threat: Fake Accounts and Identity Fraud

The company's customer review site is hacked and false negative reviews proliferate. Fake user accounts and identity fraud involve the creation of fictitious accounts or the unauthorized use of legitimate identities, typically to mask malicious actions or falsely represent metrics.

Integrity Compromise

Perpetrators may create fake patient profiles or manipulate data to skew clinical trial results at Never Age Biotech. They can also impersonate key personnel to gain unauthorized access, spread misinformation, or misrepresent product testimonials and user feedback.

Deal Impact

The discovery of fake accounts or identity fraud can have major implications for acquisitions. For High Returns Private Equity, this could mean grappling with inflated or misrepresented user metrics, altering the valuation of Never Age Biotech. The presence of such fraudulent activities can also require additional security and verification processes, adding unforeseen costs and delays to the acquisition timeline. These frauds can diminish trust if unchecked, leading High Returns Private Equity to reconsider the acquisition's terms or viability.

The Elon Musk-Twitter acquisition saga of 2022 highlights the importance of accurately reporting user metrics. Musk's apprehensions about underreported bot accounts on Twitter led to him suspending the $44 billion acquisition and threatening to abandon the deal. This indicates how cybersecurity threats amplify risks during the deal process.

Summary

The deal threat scenarios and real-world cybersecurity incidents presented in this chapter underscore the numerous vectors of cybersecurity threats that M&A transactions currently face. Controlling all aspects of the M&A transaction is only possible if cybersecurity risk is managed before, during, and after the merger or acquisition.

Chapter 3

Asymmetric and Exponential Risk

During my tenure with a renowned global technology company, I encountered an insidious cyber threat, which reinforced the importance of thorough cybersecurity due diligence during mergers and acquisitions.

A top-tier financial institution was poised to acquire a regional bank to expand its reach and solidify its market position. The institution sought our business continuity services to ensure a seamless transition and integration of operations. This step was vital in assessing the resilience of their combined technology infrastructure.

To mitigate any risks that could have heightened consequences due to the merger, I took the initiative to conduct a comprehensive due diligence security assessment of our infrastructure. The main objective was to ensure that our testing environment was fully functional and highly secure.

Several days into our evaluation, we discovered something unexpected. Hidden on a server within our infrastructure was a rootkit, a suite of software hacking tools that could grant unauthorized access and control to an outsider. In a routine scenario, while serious, such a discovery would mean addressing the threat and containing any damage, including performing relevant forensics, root cause analysis, and security improvements. The episode would have primarily financial implications limited to the costs of containment and mitigation.

However, this was not a routine scenario. The backdrop of a merger magnified the potential consequences of this discovery. The fallout would have been colossal had this rootkit been exploited, especially during the phase when the banks' data resided on our systems for testing. Beyond immediate data breaches or technical interruptions, the credibility of the two banks could be at stake. Trust, which is the foundation of all financial institutions, would be shaken. The economic ramifications of such a breach would escalate quickly, resulting in monetary losses. Even worse, the merger, a multibillion-dollar endeavor, could have collapsed.

This episode was a sober reminder that cybersecurity due diligence is not just a checkmark in mergers and acquisitions. It is the fine line that separates success from catastrophe.

Asymmetric Risk in M&A Cybersecurity

When examining risks, particularly in cybersecurity, professionals weigh factors such as the probability of threats, vulnerabilities, the likelihood that an event will occur, and the severity of the impact. Some threats may have a high probability of occurrence but lead only to minimal disruption. On the contrary, others may have a low probability of occurrence, but they would have catastrophic consequences if they did.

A cybersecurity program during regular business operations balances these factors, creating a "symmetric" risk profile. That is, there is a correlation between the probability that a threat occurs and the magnitude of its impact. Responsible companies model what-if scenarios for cybersecurity threats that exploit specific vulnerabilities and estimate the impact. Models help companies understand the correlation and symmetry between threat and impact so they can implement mitigation strategies.

However, the M&A process disrupts this symmetry and replaces it with an asymmetric risk profile. In M&A, seemingly minor oversights or vulnerabilities can lead to disproportionately severe consequences. A single cybersecurity incident can cast doubts on the integrity of the merging or acquired companies' systems, devaluing the deal by wide margins ranging from tens to hundreds of millions of dollars, as seen with the Verizon/Yahoo transaction.

The stakes of the principals are at their highest during mergers or acquisitions, and the consequences of cybersecurity lapses are magnified beyond the immediate operational implications. This asymmetric risk profile underscores the criticality of competent cybersecurity due diligence during the M&A process.

The following are six reasons why cybersecurity due diligence is essential to mitigate the asymmetric risks of mergers or acquisitions.

1. Increased Stakes

During M&A, financial, operational, and reputational risks increase largely because the value of the deal is often several times the revenue of the acquired company. A cybersecurity incident during this period can put more than just immediate financial metrics at risk; it can also jeopardize the long-term strategic goals of mergers or acquisitions.

In 2014, Marriott International announced its acquisition of Starwood Hotels & Resorts. Four years later, Marriott revealed that Starwood's guest reservation database was compromised in an earlier breach. The breach exposed the personal information of about 500 million guests. The breach, which had started in 2014 before the announcement of the acquisition, became one of the most damaging data breaches ever. Although the deal ultimately proceeded, the incident drew considerable negative attention and resulted in multiple legal actions and regulatory scrutiny. This episode highlights the complexities and reputational risks that can arise when cybersecurity issues coincide with M&A activities.

---

2. Magnified Repercussions

Minor data breaches or system intrusions can cause limited damage during regular business operations. However, the same incident during an M&A can have amplified consequences. Such events can lead to financial setbacks, erode trust, introduce delays, or even provide grounds for renegotiating deal terms.

In 2011, the EMC Corporation RSA Security division experienced a cyberattack that compromised its two-factor authentication products. Later that year, when EMC was in talks to acquire Dell, this breach became a point of concern. The incident revealed vulnerabilities in RSA's products, raising questions about the deal's valuation and the overall security posture of the merged entity. Although the deal eventually moved forward, the cybersecurity incident illustrated the increased scrutiny and implications such breaches can introduce during M&A negotiations.

3. Operational Complexities

M&A activities often involve integrating disparate IT systems, aligning different cybersecurity cultures, and granting access for due diligence purposes. These complexities can introduce vulnerabilities that do not exist during business-as-usual operations, which exposes the environment to security compromise.

In 2011, Hewlett-Packard's (HP) acquisition of Autonomy for more than $10 billion highlighted the pitfalls of operational complexity. Their underlying incompatibilities were apparent as HP sought to integrate Autonomy's systems into its own. Although not a cyber breach in the conventional sense, the difficulties arising from misaligned systems and inadequate due diligence culminated in an $8.8 billion write-down for HP. This case is a stark reminder of the increased risks that operational and cybersecurity intricacies can pose during the M&A process.

4. Reputational Impact

Mergers, acquisitions, and investment negotiations are invariably in the spotlight, attracting attention from the media, competitors, and broader markets. A cybersecurity incident during such moments can inflict reputational damage disproportionate to the incident itself, with effects that may persist for an extended period.

When Uber was in advanced investment discussions with SoftBank in 2017, the rideshare company revealed a data breach that affected 57 million users. Criticism over Uber's response to the incident amplified concerns about the company's governance and security practices. This negative publicity and increased scrutiny influenced the dynamics of the deal negotiations. Consequently, SoftBank benefited from the situation, securing a large discount and investing in Uber at a roughly 30% markdown on the initial proposed valuation. This example underscores how cybersecurity lapses can dramatically alter a firm's reputation and financial terms during high-stakes transactions.

5. Regulatory & Legal Consequences

Mergers and acquisitions inherently attract the attention of regulatory authorities. In the event of a cybersecurity breach during such transactions, companies face immediate regulatory fines and a looming threat of post-acquisition litigation. This additional scrutiny amplifies the complexities and financial implications of the deal.

The merger between Starwood and Marriott serves as a stark example of such consequences. After their data breach incident became known, Marriott faced stringent regulatory action, culminating in a $23.8 million fine levied by UK authorities for violating the European data privacy regulations officially known as the General Data Protection Regulation (GDPR). This penalty underscores the increased regulatory and legal risks that companies face during M&A transitions.

6. Executive Priorities

As M&A proceedings advance, executive leadership and senior management focus on operational integration, strategic alignment, and the financial intricacies of the deal. While these aspects of transactions are of utmost importance, they can sometimes overshadow the regular attention required for cybersecurity, increasing the risk of vulnerabilities and threats going unnoticed.

The Microsoft-LinkedIn acquisition scenario in 2016 serves as a pertinent case in point. As acquisition talks for the approximately $26.2 billion deal were in full swing, a data breach from 2012 that LinkedIn had previously addressed and believed to be contained resurfaced. However, this time the extent of the breach was much more extensive than originally reported. This resurfacing, when the leadership of both companies was deeply involved in a high-stakes merger, underscores the risks of an executive change in priorities. It emphasizes how latent issues can materialize when top leadership's attention is overwhelmingly directed toward M&A endeavors leading to inadvertent but potentially dangerous cybersecurity oversight.

Exponential Risk in M&A Cybersecurity

The merging of two distinct entities is complex, and cybersecurity risks can escalate rapidly, multiplying rather than simply adding up. A few years ago, I had the privilege of advising an international financial institution that was in the process of acquiring a leading US financial firm. These two companies were well established in their respective fields and had amassed technology assets worth hundreds of millions of dollars, with their businesses valued in the billions. However, the biggest hurdle in this acquisition was not simply combining their assets, but the differences in their cybersecurity strategies that needed to be addressed.

Each institution had customized its cybersecurity program based on its unique business needs. One had moved forward, adopting cloud technologies and the security automation benefits it offers, while the other company maintained a more traditional infrastructure. The challenge of integrating cloud systems with the on-premises systems of two merging companies is multilayered and complex from a security standpoint. Cyber risk could rapidly escalate if exploited vulnerabilities in either or both environments affected the combined entity.

Integrating their operations was always going to be a challenge due to the differences in technology management tools between the two companies. To make matters worse, we discovered that one third of the technology used by the institution being acquired was outdated and no longer supported by the vendors. This discovery was a concern because unsupported end-of-life technology is more vulnerable to cyberattacks.

The structure of the companies' security operations also presented challenges. As aforementioned, each had developed its security measures tailored to its needs. There were areas of overlap, but there were also gaps. And in cybersecurity, even the smallest gap can be a serious vulnerability.

My job was to collaborate with the stakeholders and craft a tactical cybersecurity roadmap to guide these institutions through the acquisition process. Doing so involved thorough assessments, audits, security architecture reviews, and measures to identify and patch vulnerabilities. The integration team could not simply integrate the out-of-date systems with the acquirer's systems—which processed billions of customer financial transactions daily—without introducing incalculable risk exposure. A phased, upgrade-and-integrate approach prevailed. The applied security controls provided the assurance, confirmed by independent security assessments, that the combined entity was resilient against cyber threats.

One key takeaway from this experience is that when companies with substantial digital assets merge, the cybersecurity risks can be far greater than the sum of their parts. As the frequency of mergers and acquisitions continues to increase, it is prudent to prioritize security preparation and diligence. Even seemingly minor weaknesses can result in risk exposure for both parties.

The term exponential describes a pattern of increase in which the growth rate is not additive but multiplicative. This growth pattern shows rapid nonlinear escalation over time. Applied to M&A, cybersecurity issues arising during the transaction or post-transaction can rapidly escalate damage and have far more impact than during business-as-usual times. Instead of adding risks from each party, the merged entity may have vulnerabilities that might have remained latent or undetected if the companies continued operating independently. These factors multiply the risk exposure. The following are six reasons why cybersecurity due diligence is critical to mitigating the exponential risks that arise in a merger or acquisition.

1. Multiplied Vulnerabilities

When companies merge, they do not simply add two sets of systems and data; they create a more complex ecosystem, where the interaction between old and new systems can introduce new vulnerabilities or amplify existing ones.

Consider the mammoth acquisition of 21st Century Fox by Disney in 2019. Beyond the headline financial figures, the merger meant integrating vast media libraries, user databases, digital platforms, and more. Before the merger, each entity had its cybersecurity posture, unique set of processes, and unidentified vulnerabilities. Merging these vast digital ecosystems is more than just doubling the digital assets. It also means multiplying the security challenges. Even without a direct breach scenario in this case, the intricacies of such integrations highlight the exponential risks involved in M&A cybersecurity.

2. Rapid Changes

Mergers and acquisitions often set the stage for rapid operational change. Regardless of the integration of various IT systems, the assimilation of different company cultures or procedural changes, such as employee transitions, can introduce new security vulnerabilities. A hasty or ill-prepared transition may widen the window of opportunity for cyber adversaries, intensifying the risk of each operational change.

The Dell-EMC merger in 2016 is emblematic of this reality. Dell, a giant in the personal computer space, and EMC, a leader in storage solutions, had fundamentally different operating models and cybersecurity postures. Beyond the obvious corporate cultural integration challenges and calibration of services and product lines lay the monumental task of ensuring that the layers of changes happening in real-time did not introduce exploitable vulnerabilities.

3. Different Cultures

If Company A has a robust cybersecurity training program, but Company B does not, post-merger, untrained employees from Company B can exponentially increase the risk of incidents due to human error or oversight.

4. Increased Visibility

High-profile M&As often draw attention from the media and industry. This increased visibility can lead to a surge in cyber-attacks as malicious actors see opportunities to exploit transitions.

An example is Amazon's acquisition of Whole Foods in 2017, which at the time attracted immense media and industry attention. This increased visibility can make companies involved in mergers and acquisitions targets for cybercriminals hoping to exploit transition-related vulnerabilities or gain access to consumer or proprietary data.

---

5. Interconnected Dependencies

As companies integrate, their systems become more interdependent. Exposure or a breach in one area can rapidly cascade through these interconnections, leading to widespread disruptions. For instance, an issue in a minor subsidiary's network could compromise primary operations if the networks are interconnected without proper segmentation or adequate network security controls.

In 2014, Home Depot suffered a breach in which attackers used credentials from a third-party vendor to infiltrate the main network. This highlights the risks of interconnected systems, where a vulnerability in a less secure, connected entity can jeopardize the primary organization.

---

6. Regulatory Landscape

M&A sometimes involves entities from different jurisdictions or geographies, each with a unique regulatory regime and cybersecurity-related mandate. Noncompliance or oversight in any one area during an acquisition or merger can have a ripple effect, leading to multiple regulatory breaches, possible monetary fines, and loss of deal value.

When Marriott International acquired Starwood Hotels & Resorts in 2016, they inherited a data breach that affected Starwood's reservation database. This breach led to regulatory repercussions in multiple countries and tens of millions of dollars in monetary fines. This incident is emblematic of the cybersecurity challenges faced when navigating a multi-jurisdictional regulatory landscape, especially in cross-border M&As.

Chapter 4

Engagement

Mergers and acquisitions involve a series of engagements that require the exchange of sensitive information, negotiations, and deliberations with at least two principals and facilitators, including lawyers, accountants, brokers, and M&A specialists. Including elements relevant to cybersecurity from the start is a good first step to mitigate risk during the transaction process. This chapter examines the stages of the M&A process and emphasizes the importance of implementing cybersecurity guardrails from the very beginning.

Initial Discussions

The initial interactions between buyers and sellers set the tone of the transaction process. It is a phase of exploration, assessment of shared interests, and mutual evaluation. Although the focus may be on business synergies and financial metrics, the parties should initiate good cybersecurity practices. The areas of focus are as follows.

Communication Channels

All substantive, initial, and subsequent discussions should be conducted over secure and encrypted channels, whether they are voice calls, video conferences, emails, or text exchanges. Secure communication channels ensure sensitive preliminary, pre-Letter of Intent, information remains confidential and is not exposed to eavesdropping or interception.

Non-Disclosure Agreements (NDA)

The Non-Disclosure Agreement is a document that the buyer signs agreeing to not disclose and keep confidential all information shared by the seller, such as financial statements and customer contracts. The NDA is often one of the first formal documents exchanged between parties in an M&A transaction. This protects the confidentiality of the seller's information. An NDA should include relevant cybersecurity clauses to establish a proper foundation for information protection. Stakeholders should consider cybersecurity elements such as, but not limited to, the following:

- Data Protection Clauses: The NDA must clearly define confidential information, including data related to cybersecurity programs, vulnerabilities, and security incidents. Specific provisions should be in place to manage, store, and eventually dispose of such data.

- Duration of Confidentiality: Specifying how long the information must remain confidential ensures that the security considerations are not eroded over time.

- Permissible Disclosures: In the event of a cybersecurity incident, there may be regulatory or legal obligations to disclose certain information. The NDA should outline scenarios in which such disclosures are permissible, unless legally mandated, ensuring that both parties are aligned in their understanding.

Confidential Information Memorandum (CIM)

As its name implies, the Confidential Information Memorandum (CIM) is a document that the seller, investment bank, or broker prepares, and buyers with whom the document is shared are often legally bound to keep it and the information confidential. The CIM presents the company in a positive light to acquirers. It usually includes pertinent information on the business of a company, historical financial performance, market position, and future growth prospects. Typically, the recipient must sign an NDA before the CIM is shared. Cybersecurity elements such as, but not limited to, the following should be considered:

- Secure Data Room: These document-sharing and management systems have features such as encryption of data in transit and at rest, document watermarking, and document access auditing, which provide the necessary security for deal documents.

- End-point Security: The parties should implement security controls, such as antivirus and intrusion detection software, on their local devices, such as laptops and mobile phones, to ensure that the CIM and other documents shared by the seller remain confidential and accessible only to authorized parties.

- Multifactor Authentication: Password-protected data rooms are relatively easy to breach by skilled hackers. Adding a layer of security with multifactor authentication, such as texting a random numeric code for the user to enter along with their password, reduces the chances of access and the likelihood of compromise.

Letter of Intent (LOI)

Signing the LOI by the buyer and the seller signifies a mutual interest in proceeding with the transaction. Although it may not legally bind the parties to perform a transaction, it is an essential document that sets out the primary terms and conditions of the deal. The LOI precedes the signing of a definitive agreement to purchase a company or its assets. A period is reserved after the LOI is signed but before the definitive agreement to conduct financial, legal, and other due diligence. From a cybersecurity program management perspective, this is a short timeframe, perhaps 30 to 90 days. When considering cybersecurity risks to the transaction, this compressed timeframe requires the efficiency of a prioritized approach to evaluate cybersecurity controls.

The LOI should establish preliminary expectations for cybersecurity due diligence. This could consist of security certifications, documentation, assessments, interviews, and logs. Chapter 5 introduces a comprehensive framework for managing cybersecurity due diligence.

The Definitive Agreement

The Definitive Agreement (DA) is enhanced by information from a thorough cybersecurity assessment performed during due diligence. It informs various aspects of the agreement, ensuring a sound and secure transaction. The following are the cybersecurity-related clauses that may be in the DA. This is not an exhaustive list of clauses or legal advice. See the legal disclaimer at the end of this section.

- Valuation Adjustments: If cybersecurity vulnerabilities or previous incidents are discovered that could pose financial or reputational risks, the acquirer may adjust the valuation and purchase price.
- Indemnification Clauses: Indemnifications may be tailored based on the cybersecurity posture of the seller, providing financial protection for post-acquisition liabilities linked to pre-existing cybersecurity issues.
- Representations and Warranties: Specific clauses related to the accuracy, completeness, and functionality of the cybersecurity practices, policies, and infrastructure of the target company.
- Remediation Commitments: The seller might be required to address specific cybersecurity issues before the deal is finalized, or the buyer might seek to decrease the purchase price to accommodate the cost of addressing these issues post-acquisition.
- Regulatory Compliance: Ensure that the target company adheres to applicable data protection laws and industry-specific cybersecurity regulations.
- Data Handling and Protection: Commitments to protect sensitive data during and after the acquisition, ensuring that no unauthorized access or data leakage occurs.
- Employee Training and Culture: Ensure that cybersecurity training and a security-conscious culture are intact and will be maintained or integrated post-acquisition.

Role of Cyber Insurance

Although cyber insurance plays a role in mitigating the financial impacts of a cyberattack, it does not protect against loss of customer confidence, business reputation, or legal standing, and it may not protect against deal implosion. It applies after an adverse security event, providing a safety net, but is not a preventative measure. However, maintaining cyber insurance is a good business risk management practice.

Summary

M&A deals involve immense investments in terms of financial and strategic value. Cybersecurity incidents can profoundly affect the perceived and actual value of a deal.

- Deal Integrity: By setting clear expectations early on, both parties can mitigate the risk of unexpected revelations or cyber incidents during due diligence or post-acquisition that could jeopardize the deal.
- Negotiations: The talks can proceed with a shared understanding, reducing any friction points related to cybersecurity.
- Trust: Transparency and clarity on cybersecurity matters can boost mutual trust and foster the collaborative environment that is essential for smooth M&A transactions.

Legal Disclaimer

The preceding section does not provide legal advice. M&A transactions typically involve law firms or attorneys who represent buyers and sellers. Companies involved in M&A should seek legal advice on the cybersecurity provisions included in all transaction documents.

Chapter 5

Mergers and Acquisitions Cybersecurity Framework

Unmitigated cybersecurity risks during M&A transactions persist after the transactions. The Mergers and Acquisitions Cybersecurity Framework (MAC-F) provides stakeholders with a structured and comprehensive approach to mitigating cybersecurity risk initially, during, and after transactions. Each M&A transaction has unique attributes; the MAC-F adapts to different deal structures, including asset and share purchase/sale scenarios. The MAC-F is adaptable to entities of all sizes, whether they are classified as small, medium, or large companies.

The framework emerged from my first-hand experience and is tailored to companies preparing for acquisitions/sales or mergers. The primary objective is to mitigate cybersecurity risks that may derail the transaction, protect against decreases in valuation after due diligence, and mitigate reputational harm caused by cyber incidents.

For the seller, this also minimizes the chance of post-transaction financial holdbacks due to cybersecurity challenges. This prioritized approach was successfully leveraged and adapted to companies in various industries. These include firms with multimillion to multibillion-dollar valuations spanning various sectors, including global financial institutions, technology giants, midmarket manufacturing, retail, e-commerce, cryptocurrency, blockchain, and Artificial Intelligence companies.

A foundational aspect of this approach is that it is based on more than a decade of experience working with companies to build robust, mature cybersecurity programs and preparing them to consistently succeed in reducing enterprise cyber risk and audits based on regulations and standards, such as the Federal Financial Institutions Examinations Council (FFIEC) Security Handbook, American Institute of Certified Public Accountants (AICPA) SOC I and SOC 2, Payment Card Industry Data Security Standard (PCI-DSS), and ISO/IEC 27001 security standards.

Companies taking this approach have experienced negligible security findings. Their clear cybersecurity due diligence records have facilitated smoother transactions and preserved tens of billions of dollars in cumulative transaction value.

The MAC-F addresses cyber threats to M&A in five categories. This methodology provides a means to estimate the capital and recurring costs associated with risks and risk mitigation. The MAC-F gives M&A stakeholders clarity regarding cybersecurity risk exposure and transparency regarding costs that can be factored into deal negotiations. Companies seeking a sale should execute the MAC-F before signing the LOI in preparation for the buyer's due diligence.

Companies pursuing a merger or acquisition would execute the MAC-F at the beginning of a formal due diligence phase.

There are five categories of cybersecurity risk controls for the MAC-F.

1. Prevent

This control category comprises the security domains of Asset and Configuration Management, Data Protection, User Account and Identity Management, Network Security, System Security, Endpoint Security, Application Security, and Physical Security. These controls prevent cybersecurity threats from compromising the transaction and the merged or acquired entity after the transaction.

2. Detect

This control category comprises Monitoring, Vulnerability Management, and Past Security Incidents. These controls detect cybersecurity threats, vulnerabilities, and unmitigated past security incidents before they become risks that compromise M&A transactions.

3. Respond

This control category comprises the security domains of Business Continuity and Disaster Recovery (BC/DR), Incident Management, and Security Personnel. These controls provide a means of responding to cybersecurity incidents that can occur during and after a merger or acquisition.

4. Manage

This control category comprises the security domains of Security Documentation and Governance, Management and Board Focus, and Organizational Awareness. These controls provide corporate oversight, human participation, and processes necessary for effective cybersecurity management. Deficiencies in this control area increase risk during and after M&A transactions.

5. Comply

This control category comprises Customer Contracts, Audits and Regulations, and Vendor Management security domains. These controls ensure compliance with customers' contractual security clauses, audits, and legal and regulatory mandates and ensure that M&A transactions are not impacted by incidents of non-compliance.

The MAC-F provides a prioritized methodology to identify cyber risks throughout the M&A lifecycle. Its five control categories—Prevent, Detect, Respond, Manage, and Comply—facilitate multidimensional due diligence tailored to the nuances of complex transactions.

Prevent controls focus on protecting assets, data, systems, applications, and infrastructure by "closing doors" to threats before they can cause harm. Comprising technological and policy controls, Prevent aims to protect the environment against compromise. The technical controls include firewalls, access management, endpoint security, and encryption. Administrative controls encompass security policies, standards, procedures, and training to build employee resilience to threats.

Prevent addresses risks from threat actors who may exploit vulnerabilities to steal intellectual property or commit other compromises before, during, and after the transaction.

Detect controls provide visibility into threats that may penetrate defenses by enabling continuous security monitoring, behavior analytics, simulated attacks, and vulnerability assessments that highlight remediation needs.

Shortcomings in Detect can allow advanced persistent threat actors to dwell undetected within networks for extended periods, whether they entered through overlooked preventative controls or evaded them. As an illustration, a nation-state-backed threat group can leverage stealthy, custom-made malware to slowly siphon proprietary formulas. Right before closure, they could threaten to leak the stolen intellectual property if a ransom is not paid. The lack of robust threat detection and visibility introduces uncertainty, which can derail transactions.

Respond capabilities allow for rapid containment, eradication, and recovery when incidents occur, using playbooks, skilled personnel, and incident respond processes. An effective response prevents isolated incidents from becoming organization-spanning crises.

Respond gaps can lead to cascading impacts when sophisticated attacks manifest themselves in compromised networks. For example, deficient incident response planning may allow ransomware with self-propagation functions to spread across uncontained systems. This could cause widespread destructive shutdowns, resulting in millions in lost revenues and business disruptions during the M&A transition. Such major disruptions introduce substantial uncertainty into financial valuations and synergy projections.

Manage controls provide a governance foundation and resources for sustaining effective security programs with diligent oversight.

Metrics offer visibility for performance. Training strengthens human resistance to risks.

Poor management can allow preventable risks to persist and be inherited when companies merge. For instance, employees retaining bad security habits post-close, such as password reuse, could enable insider theft of intellectual property. Security issues like these can undermine the anticipated synergies.

Comply capabilities ensure adherence to contractual security requirements, industry regulations, and customer commitment. Violations introduce legal, financial, and reputational risks.

Comply deficiencies may result in substantial fines, lawsuits, and public recrimination if data breaches or noncompliance issues are uncovered. For example, non-compliance with GDPR requirements can lead to steep legal penalties. Non-compliance incidents can severely damage brand value and customer trust.

The five risk categories of the MAC-F—Prevent, Detect, Respond, Manage, and Comply—function as a strategic lens for stakeholders by focusing on distinct aspects of cybersecurity tailored to M&A deals, allowing targeted identification of specific gaps by security function.

Security Domains: Risk Focus

The methodology has risk categories that contain specific security domains, such as Asset Management and Incident Management, which address risk within that category. This categorization is not only taxonomical but provides a focus for precise risk assessment. The security domain approach is well supported by widely accepted standards for security risk management, such as NIST 800-53, and ensures that risk mitigation efforts are precisely targeted and not diffused. However, the MAC-F includes additional security domains, such as Customer Contracts and Past Incidents, which are high-risk factors that can harm M&A deals.

Diligence Evidence: A Multifaceted Approach

Collecting evidence is essential to conduct cyber due diligence. The MAC-F due diligence evidence collection from each security domain is a process that involves various assessment tasks, each crafted to comprehensively evaluate security measures.

Policies and Procedures: This subtask involves a careful review of all documentation related to security policies, standards, guidelines, and procedures, examining the formal policies established for each domain, and ensuring that they are comprehensive and adhered to by the organization.

System Configurations: This involves auditing and assessing the technical controls and system configurations implemented for each domain, reviewing firewall rules, access controls, system logs, and more, ensuring that systems are not merely compliant with policies but are configured for optimal security.

Interviews: This involves conducting inquiries through questionnaires and discussions with key personnel, such as CISOs, security engineers, and business process owners, to gain insight into security practices, ensuring that policies and configurations are not just theoretically sound but are also practically applied.

Logs: This involves compiling and examining security logs, audit trails, monitoring outputs, and other historical security data to identify issues, ensuring that past incidents are learned from and that future incidents can be predicted and prevented.

Additional Information: This involves looking for any other artifacts, processes, or evidence that may clarify the security posture for each domain.

Deal Impact: Quantifying the Immediate Risk

The methodology then progresses to assess the deal impact, quantifying the likely impact of the identified security risk on the M&A deal and post-closing if it remains unmitigated. This assessment is not merely technical but also considers how vulnerabilities could affect the deal's valuation, due diligence, negotiations, or legal aspects. It ensures that cybersecurity risk is translated into business risk and provides a clear and compelling case for mitigation.

Post Impact: Mitigating Post-Transaction Risk

Post-impact risk is then evaluated by examining the prospective impact of the security risk after the deal closes and integration begins, whether the companies have combined or it is an outright acquisition. This involves examining the downstream impacts on operations, customers, revenues, and other business aspects. It ensures that the methodology is focused on the present and anticipates, identifies, and mitigates risks that may persist after the deal closes.

Mitigation: Risk Reduction

Mitigation strategies were formulated to capture the measures recommended to address identified security gaps and weaknesses to reduce risk exposure. These mitigations are proactive and aim to strengthen defenses and improve the target's security posture. They ensure that the organization can achieve the value that stakeholders envisioned before initiating the transaction.

Capital and Recurring Cost: Financial Transparency

The MAC-F methodology ensures that the recommended risk mitigation activities are financially transparent by estimating the costs required to implement and sustain them. This includes expenses such as technology purchases, process redesign, and other nonrecurring charges. Annual costs include staffing, licenses, subscriptions, maintenance, and operating expenses. This approach provides M&A stakeholders with an unparalleled approach to understanding and managing cybersecurity risk, its impact on transaction viability and valuation, and its post-transaction financial and strategic value.

The MAC-F methodology offers a structured, thorough, and principled approach to identifying, evaluating, and mitigating cybersecurity risks throughout the M&A lifecycle. The following six chapters examine how the specific security domains within the MAC-F function to mitigate threats that might disrupt M&A transactions.

Chapter 6

MAC-F: Prevent

Asset Management

Asset management is a domain that demands attention. It involves systematically tracking, evaluating, and protecting the digital and physical assets of an organization, ensuring that they are optimally used, protected, and in compliance with regulatory mandates. Sellers and buyers must prioritize effective asset management, ensuring that the organization's assets are well-documented, utilized, and safeguarded against cybersecurity threats.

Action: In the lead-up to M&A activities, sellers should prioritize performing a comprehensive review and documentation of all assets, ensuring that they are optimally utilized, protected, and compliant with all regulatory mandates. This involves tracking each asset's lifecycle and ensuring that they are maintained, patched, and protected against threats. During their due diligence, buyers must validate the accuracy and completeness of the seller's asset management practices. They should validate the inventory, ensure it is protected, and that any serious vulnerabilities have a remediation plan.

Benefits: Demonstrating a robust asset management program assures buyers of the organization's operational efficiency. It safeguards the organization against risk of compromise to unknown or unmanaged assets. For buyers, validation of the seller's asset management practices ensures that they do not inherit any related risks or exposures and that the information and physical assets are substantiated and protected from compromise.

Deal Impact: Buyers should consider the efficacy and robustness of the seller's asset management practices during deal negotiations. This could involve evaluating the management of the life cycle of the assets, putting in place safeguards to protect them against threats, and complying with relevant regulatory mandates. Any gaps or inefficiencies in the seller's asset management practices could require adjustments in the value or terms of the deal, ensuring that the buyer does not assume undue risk or inherit a weakened operational posture.

Artifacts: Key artifacts to validate asset management practices in an organization include an Asset Management Policy, which outlines the organization's approach to tracking, utilizing, and protecting assets. Asset lifecycle reports provide a detailed overview of the lifecycle of each asset, ensuring that it is maintained, patched, and protected.

In addition, Compliance Reports validate the compliance of each asset with regulatory mandates, and Vulnerability Assessment Reports provide information on its security posture. These documents provide a comprehensive overview of the organization's asset management practices, ensuring that they are optimally utilized, protected, and compliant.

Summary

- Rationale: Effective asset management ensures optimal utilization, protection against threats, and compliance with regulatory mandates.

- Insight: A robust asset management program provides operational efficiency and protection against cybersecurity threats and vulnerabilities.
- Deal Impact: Inefficiencies or vulnerabilities in asset management can introduce risks and may require adjustments in the valuation or terms of the deal.

User Accounts & Identity Management

Understanding and ensuring the integrity of user accounts and identity management systems is fundamental to cybersecurity management and should not be overlooked in a company merger or acquisition. Buyers and sellers must prioritize this domain because legitimate users are the lifeblood of business.

Action: Both parties should ensure the integrity and security of user accounts and identity management systems. This involves a thorough review and, if necessary, the implementation of robust identity management systems to ensure that only authorized entities can access resources. Implementing multifactor authentication, single sign-on solutions, and conducting periodic audits of user accounts and access rights are essential to identify and rectify discrepancies and protect access to sensitive resources.

Benefits: Effective user account and identity management protects a company's digital assets from unauthorized access and use and is a core component of best practice security management. Ensuring robust user access controls and deploying identity management solutions reduces the likelihood of insider threats and unauthorized access. Robust user and identity management protects assets and reduces the likelihood of cybersecurity incidents during all phases of M&A transactions.

Deal Impact: Buyers should examine the target's identity management systems, including measures to detect fake or compromised accounts, processes to rectify discrepancies, and the scalability of these systems as the user base grows. If evaluations uncover gaps or weaknesses, the costs of overhauling or improving these systems should be considered and factored in during deal negotiations and may influence the transaction terms or price.

Artifacts: Proper documentation is necessary to prove the effectiveness of user account and identity management. This includes the User Access and Identity Management Policy, which outlines the company's security principles for managing access, and the User Access Review Report, which provides a snapshot of user permissions and recent changes. Identity Management configuration documents and Audit Logs provide insight into user access and permissions, offering a comprehensive view of the company's approach to securely managing user identities and access.

Summary

- Rationale: Authentic users and account validity are a core requirement for digital platforms, and identity management is mandatory for preventing unauthorized access to corporate systems.
- Insight: Fake accounts can distort user metrics, affecting platform credibility and revenue streams, whereas unauthorized users pose a threat to system security.
- Deal Impact: A skewed user account base may misrepresent a company's value and impact post-acquisition projections, with unauthorized users presenting a latent threat within the systems of a company.

Data Protection and Encryption

Strong data protection and encryption is not negotiable in M&A. Companies must prioritize protecting sensitive and proprietary information. This includes all types of sensitive information including that of its employees, customers, vendors and partners. Additionally, companies may be subject to data protection regulations such as the European General Data Protection Regulation (GDPR) which imposes financial penalties for non-compliance.

Action: Sellers must prioritize the implementation and validation of robust data protection and encryption measures. They should thoroughly assess data storage, data transmission, and access, and ensure that encryption protocols for data at rest and in transit are implemented according to the industry approved, current standards. Buyers should rigorously assess the suitability and effectiveness of a target's data protection measures and be prepared to factor in the costs and resources required for improvements if the controls and protocols are deficient.

Benefits: Demonstrating robust data protection measures can enhance appeal during M&A negotiations and reduce the risk of data breaches, and provide assurance of compliance with regulatory requirements. For buyers, ensuring that effective data protection measures are in place is a deal risk mitigation strategy and investment in the long-term value of the acquisition.

Deal Impact: Buyers should be vigilant about the need for post-acquisition remediation and costs, which could involve hold-back clauses or other adjustments to the transaction's terms or price. Ensuring that the target company adheres to data protection mandates is advised to avoid financial and other penalties for non-compliance.

Artifacts: Tangible documentation validating commitment to data protection and encryption should be obtained. Key artifacts may include a Data Protection Policy, Encryption Protocols and Standards documents, and Encryption and Encryption Key Management documentation, providing a transparent and comprehensive view of the company's data protection and encryption practices.

Summary

- Rationale: Data protection safeguards privacy, ensures information confidentiality, and maintains regulatory compliance.
- Insight: Adherence to data protection mandates such as GDPR is required to avoid penalties and ensure sustained regulatory compliance.
- Deal Impact: Lapses in data protection can trigger legal consequences, erode customer confidence, and harm a business's reputation.

Network Security

Network security is essential for protection against cyber threats and unauthorized access. Ensuring the integrity, confidentiality, and availability of data and communications is fundamental to operational continuity and the protection of sensitive information. The cybersecurity posture of a target company can be evaluated through penetration tests, vulnerability assessments, and network security audits.

Action: In anticipation of M&A activities, sellers should prioritize the assessment and fortification of network security controls. This includes the implementation of robust perimeter defenses, ensuring network segmentation, deploying intrusion detection systems, and monitoring network traffic. The use of encryption for network traffic and the security of remote access is an absolute necessity for securing communications. Additionally, sellers must assess exposure points, deactivate unnecessary services, and establish a systematic patching process to address vulnerabilities.

Benefits: Strong network security controls indicate a commitment to good cybersecurity practices and intrinsically protects a company's value. Effective network controls—such as firewalls, intrusion detection systems, and network segmentation—function as barriers against threats, safeguard sensitive systems, and prevent the lateral movement of threats across the network. Employing encryption protects data in transit, while a systematic approach to patching and service management mitigates exploitation opportunities.

Deal Impact: Evaluation of target company's network security practices and incident history is a prudent step for acquirers. Addressing network security issues, vulnerabilities, or breaches can be resource-intensive and may require major post-acquisition investments. Consequently, findings from network security evaluations may be leveraged to negotiate adjustments in the acquisition price or terms, accounting for any liabilities or remediation costs that may be incurred post-acquisition.

Artifacts: Key artifacts should include the Network Security Policy, which defines protections, acceptable use, and access requirements. Network Topology Diagrams provide visibility into segmentation and data flows, whereas Vulnerability Scan Reports highlight exposures and remediation recommendations. Firewall configuration standards document access controls and rules for traffic filtering. Together, these documents support the company's network security posture and help identify residual risks related to networks.

Summary

- Rationale: Understanding network security practices and incident history provides information on a target's cybersecurity maturity and risk management capabilities.

- Insight: Network security audits, incident histories, and remediation efforts are tangible measures of an organization's network security capabilities and overall cybersecurity health.

- Deal Impact: Network security findings can influence acquisition terms, valuation, and post-acquisition cybersecurity strategy, necessitating adjustments to account for liabilities or remediation costs.

Chapter 7

MAC-F: Prevent (cont.)

Systems Security

System security is fundamental to safeguarding systems against unauthorized access, disruptions, and vulnerabilities, especially in M&A. Policies, processes, and controls designed to protect system infrastructure, hardware, and software from cyber threats are necessary for evaluating the cybersecurity posture of a target company.

Action: Sellers, in preparation for M&A activities, should prioritize evaluating and fortifying system security controls. This involves implementing secure system configurations, restricting unnecessary access, ensuring prompt patching of operating systems and applications, and monitoring systems for anomalies. Multifactor authentication (MFA) should be mandated for administrative access, and sellers should also assess system interconnections and data flows to identify any pathways for lateral movement.

Benefits: Demonstrating secure system configurations and access controls signals a commitment to cyber hygiene. Ensuring that systems are patched and monitored for threats that minimize the attack surface while requiring MFA for administrative access prevents unauthorized changes. Understanding the connections between systems allows for more effective segmentation. These controls collectively limit risk exposure and protect against cyber threats during and after the transaction.

Deal Impact: For acquirers, evaluating the security practices of the target company's system and incident history is necessary for effective due diligence. Addressing system security issues, vulnerabilities, or breaches can be resource-intensive and may require large investments after an acquisition. Consequently, findings from system security evaluations can be used to negotiate adjustments in the acquisition price or terms, accounting for liabilities or remediation costs that may be incurred after acquisition.

Artifacts: Key artifacts should include System Security Standards that outline configuration baselines. System Patching Reports provide insights into the update status, whereas Access Control Matrices define appropriate user permissions. Architectural Diagrams map system interconnections, and Audit Logs demonstrate monitoring capabilities. These documents validate efforts to improve the security of the system and identify residual risks.

Summary

- Rationale: Understanding system security practices and incident history provides information on a target's cybersecurity maturity and risk management capabilities.
- Insight: System security audits, incident histories, and remediation efforts provide tangible measures of an organization's system security capabilities and overall cybersecurity health.

- Deal Impact: Systems security findings can influence acquisition terms, valuation, and post-acquisition cybersecurity strategy, necessitating adjustments to account for liabilities or remediation costs.

Endpoint Security

Endpoint security is a necessity to protect both the devices and the networks they connect to from security compromise. This involves ensuring the security of all endpoints, including user devices and back-end network connected systems.

Action: In the lead-up to M&A activities, sellers should prioritize a comprehensive review and reinforcement of endpoint security controls. They should ensure that all endpoints are secured with updated anti-malware and antivirus solutions, that robust access controls are implemented, and that all endpoint devices are patched and configured securely. Additionally, user training programs should be conducted to ensure that all employees are aware of and comply with endpoint security policies and practices.

Benefits: Ensuring robust endpoint security provides sellers with several advantages. It minimizes the risk of breaches through endpoints and is a foundational aspect of cybersecurity management. Demonstrating a commitment to endpoint security is a best practice for managing threat vectors that may be potential openings for systems and data compromise. Functioning endpoint security controls assure buyers of the company's ability to protect its network and data, thus reducing potential deal complications and liabilities.

Deal Impact: Acquirers should evaluate the seller's endpoint security practices and ensure that they align with industry best practices. This involves reviewing the seller's Endpoint Security Policy, Anti-Malware/Antivirus Deployment Records, User Training and Compliance Records, and Incident Response Logs. Buyers should also conduct independent vulnerability assessments and penetration testing to validate the seller's endpoint security posture. If gaps or vulnerabilities are identified, buyers may consider adjusting the acquisition price or terms to account for the costs and risks associated with strengthening endpoint security post-acquisition.

Artifacts: To validate a company's commitment to endpoint security, key artifacts should include the Endpoint Security Policy, which outlines the company's approach to securing endpoints. Anti-malware/Antivirus deployment records provide evidence of the technical defenses in place. User Training and Compliance Records demonstrate the company's commitment to ensuring user awareness and compliance with endpoint security policies. Incident Response Logs provide insight into a company's ability to respond to and manage endpoint security incidents. These documents validate the company's endpoint security posture and provide buyers with a comprehensive view of its approach to securing the endpoints.

Summary

- Rationale: Secure endpoints are fundamental to safeguarding the network and protecting organizational data.

- Insight: Effective endpoint security extends beyond technology deployment and includes user compliance and active threat management.

- Deal Impact: The practical strength of endpoint security directly influences the risk and valuation parameters of the acquisition, necessitating a thorough review.

Application Security

Application security is a requirement for safeguarding organizational data and maintaining stakeholder trust. It involves implementing measures and practices to protect applications from threats and vulnerabilities throughout their lifecycle, ensuring the confidentiality, integrity, and availability of data and services.

Action: In anticipation of M&A activities, sellers should prioritize the evaluation and strengthening of application security controls. This involves ensuring secure coding practices, conducting regular application security tests, and promptly patching identified vulnerabilities. Sellers should also assess the security of application deployment platforms and ensure effective code review and remediation processes are in place.

Benefits: Ensuring robust application security provides sellers with several advantages. It minimizes the risk of application breaches and underscores the commitment to reducing vulnerabilities. Furthermore, it provides buyers with assurance of the company's ability to protect its applications and data and reduces deal complications and liabilities.

Deal Impact: Acquirers should examine the seller's application security practices to ensure that they align with industry best practices and regulatory requirements. This involves reviewing the seller's application security policies, testing results, and studying the incident response and remediation records. Buyers should also consider conducting independent application security tests to validate the seller's application security posture. If gaps or vulnerabilities are identified, buyers may consider adjusting the acquisition price or terms to account for the costs and risks associated with strengthening application security post-acquisition.

Artifacts: To validate a company's commitment to application security, key artifacts should include the Application Security Policy, which outlines the company's approach to securing applications. Application Security Test Reports provide evidence of security testing and results. Application Review and Remediation Records demonstrate a company's ability to manage and correct security coding errors. These documents validate the company's application security posture and provide buyers with a comprehensive view of the approach to securing applications.

Summary

- Rationale: Understanding application security practices and coding error correction history provides insight into a target's cybersecurity maturity and risk management capabilities.

- Insight: Application security audits, code fix history, and remediation efforts are tangible measures of an organization's application security capabilities and overall cybersecurity health.

- Deal Impact: Application security findings can influence acquisition terms, valuation, and post-acquisition cybersecurity strategy, necessitating adjustments to account for liabilities or remediation costs.

Physical Security

Physical security protects the assets, premises, and personnel of an organization. It is the sum of all processes, people, and technologies deployed to protect the physical assets of the organization from unauthorized access, damage, or theft.

Action: Entities considering a sale should prioritize evaluating and strengthening physical security controls in all facilities, assets, and infrastructure. This involves the deployment of layered safeguards, such as perimeter fencing, alarms, access control systems, video surveillance, and security personnel. Sellers must also monitor access to sensitive areas, particularly those housing sensitive systems or data. They should conduct periodic physical access and control audits to identify and rectify deficiencies.

Benefits: Sellers assure buyers of their commitment to protecting their facilities and assets by demonstrating robust physical security controls. Alarms, biometric access controls, video surveillance, and door locks are examples of controls that work together to protect an organization against physical threats.

Deal Impact: Addressing physical security issues, vulnerabilities, or breaches can be resource intensive and may require post-acquisition investments. Consequently, findings from physical security evaluations may be leveraged to negotiate adjustments in the acquisition price or terms, accounting for liabilities or remediation costs that may be incurred post-acquisition.

Artifacts: Validating a company's commitment to physical security involves obtaining artifacts, such as Physical Security Policies, which outline protections and access restrictions. Facility Access Control Matrices delineating appropriate access levels and Physical Security Audit Reports that reveal control gaps and corrective actions provide evidence regarding the state of controls. Video surveillance logs provide evidence of monitoring and further validate the commitment of a company to a secure environment.

Summary

- Rationale: Understanding physical security practices and incident history provides information on the security maturity and risk management capabilities of a target.
- Insight: Physical security audits, incident histories, and remediation efforts provide tangible measures of an organization's physical security capabilities and overall security health.
- Deal Impact: Physical security findings can influence acquisition terms, valuation, and post-acquisition security strategy, necessitating adjustments to account for liabilities or remediation costs.

Chapter 8

MAC-F: Detect

Monitoring

During a merger or acquisition, cybersecurity threat and vulnerability monitoring are imperative for the early detection of threats that could impact the deal or its valuation. The monitoring of networks, systems, and applications should be prioritized by sellers and buyers so that threats can be detected before they escalate into transaction-impacting incidents.

Action: Sellers should deploy security tools and solutions to continuously monitor company networks, systems, and applications, ensuring that real-time alerts inform security personnel about anomalies or suspicious activities. This monitoring should be extended to cloud environments, third-party integrations, and remote access points, with regular logs and traffic pattern reviews to refine detection capabilities and adapt to evolving threats. Buyers must assess the adequacy and effectiveness of the seller's monitoring capabilities, ensuring that the security tools and solutions deployed provide comprehensive oversight of IT assets and network traffic. They should validate that the seller is responsive and takes action to address anomalies and suspicious activities.

Benefits: Implementing comprehensive monitoring practices assures buyers of a company's commitment to threat detection and mitigation, facilitates early detection of security incidents, and provides information on system performance, user behavior, and vulnerabilities. For buyers, ensuring that the acquisition target has robust monitoring practices safeguards the deal's integrity. It ensures long-term security and stability of the investment by protecting against future threats and vulnerabilities.

Deal Impact: Buyers should consider the robustness and ability of sellers to adapt to evolving threats. Gaps or weaknesses in monitoring mean that threats to the company may go undetected and pose a risk to the transaction and post-closing.

Artifacts: Tangible documentation that validates the commitment to monitoring is a requirement. Key artifacts may include Monitoring Policy Documents, Logs, Analysis Reports, and configuration documents for monitoring tools and solutions. These provide a transparent and comprehensive view of the company's monitoring practices and preparedness.

Summary

- Rationale: Effective monitoring ensures that an organization can detect and respond to cybersecurity threats and vulnerabilities.

- Insight: Continuous monitoring of IT assets and network traffic is necessary for early detection and mitigation of threats.

- Deal Impact: Weak monitoring practices can expose a company to cybersecurity threats, which impact the deal's valuation and feasibility.

Vulnerability Management

Vulnerability management is a performance indicator of a company's cyber maturity and overall security posture. It involves identifying, remediating, and mitigating weaknesses that cyber adversaries may exploit. Given that vulnerabilities are inevitable—arising from various sources, such as outdated software, misconfigurations, or programming defects—it is essential to have tools, processes, and people to mitigate these weaknesses consistently and programmatically. Effective vulnerability management involves continuous detection, assessment, remediation, and verification cycles. Understanding the efficacy of this cycle in a target company during an acquisition can provide deep insights into its cybersecurity culture and operational readiness.

Action: In anticipation of M&A activities, sellers should prioritize establishing a robust vulnerability management strategy. This involves regularly scanning the IT infrastructure, applications, and systems of a company to identify vulnerabilities. Once detected, a systematic approach should be implemented to assess, prioritize, and remediate these vulnerabilities based on their exploitability and impact. Integrating threat intelligence feeds to stay up to date on emerging vulnerabilities and threats, regular patching and updates, and periodic penetration tests are essential to ensure that vulnerabilities are effectively addressed.

Benefits: A comprehensive vulnerability management approach offers sellers multiple advantages. Identifying and addressing vulnerabilities assure buyers of a company's commitment to maintaining a secure environment. During the transaction phase, effective vulnerability management minimizes the likelihood of breaches or exploits, ensuring that company assets and data are protected. Consistent mitigation of vulnerabilities improves a company's security posture, signals a commitment to cybersecurity best practices, and may even make the company more attractive to certain buyers.

Deal Impact: For buyers, examining the target's vulnerability management practices is imperative. This involves understanding the frequency and depth of vulnerability assessments, patch management processes, average vulnerability closure time, and tactical response to newly discovered vulnerabilities. Discovering gaps or inefficiencies in a target's vulnerability management program presents challenges and opportunities. It provides a tangible negotiation point in the acquisition talks, where the buyer can either seek remediation before finalizing the deal or adjust the deal value to account for the post-acquisition investments needed to bolster this area.

Artifacts: Key artifacts should include a Vulnerability Management Policy document, which outlines the company's approach to detecting, assessing, and remediating vulnerabilities. Vulnerability Scan Reports and Penetration Testing Results provide a detailed account of the identified vulnerabilities and the company's remediation efforts. Patch Management Logs and Threat Intelligence Integration Reports offer insights into a company's ongoing efforts to stay updated and address emerging vulnerabilities. These documents provide buyers with a transparent and comprehensive view of the vulnerability management practices of a company.

Summary

- Rationale: Regular vulnerability scans and patches are imperative for protecting enterprises from persistent threats.
- Insight: The 2017 NotPetya attack exploited unpatched systems, causing a reported $10 billion in damage.
- Deal Impact: Weak vulnerability management is a precursor to exploits that can impact the entity during the merger or acquisition and afterward.

Past Security Incidents

Examining past security incidents, once they have been addressed and closed, is not typical in standard security programs. However, in M&A it is vital because past incidents, even those that happened years before the transaction, can have a negative impact on the deal. Examining past security incidents gives sellers and buyers insight into any latent vulnerabilities and provides insight into the effectiveness of the incident resolution process. A detailed review of historical incidents, their management, and follow-up actions is essential to understanding future cybersecurity risks. It provides insight into whether the target company has residual exposure and whether they are equipped to effectively manage security incidents.

Action: Sellers should conduct a comprehensive review of past security incidents. They must implement corrective measures to prevent them from happening again and maintain transparency about these incidents. Buyers should examine the documentation of past incidents, assess the seller's response and mitigation methods, and confirm that the identified vulnerabilities are resolved. They should understand the characteristics, frequency, consequences, and results of incidents. Additionally, buyers should assess the organization's capability to detect, address, recover from, and adapt to cyber threats.

Benefits: Addressing past security incidents with transparency benefits the organization by emphasizing corrective measures and demonstrating a mature stance on cybersecurity. This openness strengthens credibility in M&A discussions and shows a commitment to resolving issues and learning from mistakes. A comprehensive record of previous incidents and their resolutions provides insight into the organization's capabilities in incident management and business resilience, which reduces uncertainties and challenges during transactions.

Deal Impact: Buyers should assess the seller's current cybersecurity position based on past incidents. They should determine whether incidents were appropriately resolved and if they highlighted any risks. During negotiations, buyers can use this information to either request enhanced cybersecurity measures before finalizing the deal or suggest a revised deal valuation that accounts for the risk and the expense of future cybersecurity enhancements.

Artifacts: Key artifacts might include past security Incident Reports detailing incidents and their causes, impacts, and responses. Corrective Action Plans and Post Incident analysis documents shed light on mitigation actions and lessons learned. Additionally, third-party evaluations or audits conducted after an incident to verify corrective actions can demonstrate dedication to addressing incidents and improving security.

Summary

- Rationale: Historical incidents offer insight into an organization's cyber resilience and ability to respond to threats.
- Insight: The far-reaching impact of breaches, such as Yahoo, highlights the repercussions of overlooking past incidents.
- Deal Impact: Recurrent or severe past breaches can hint at embedded security flaws, resulting in post-acquisition financial and reputational liabilities.

Chapter 9

MAC-F: Respond

Business Continuity and Disaster Recovery (BC/DR)

Business Continuity and Disaster Recovery (BC/DR) plans are necessary for managing a company's resilience and operational stability and are a must during M&A. Both sellers and buyers must prioritize BC/DR to recover from disruptions and protect the company's reputation, revenue, and operational performance, especially during unexpected events, such as natural disasters, cyber incidents, or pandemics.

Action: Sellers must formulate and validate a comprehensive BC/DR plan to ensure uninterrupted operations during unexpected interruptions. This involves creating a strategic plan that includes data protection, recovery protocols, and operational contingency strategies with periodic testing and refinement to navigate emerging threats and changing business dynamics. On the other hand, buyers must rigorously assess the adequacy and effectiveness of the BC/DR preparedness of a target and its impact on the business during and after the transaction.

Benefits: For sellers, a functioning BC/DR plan amplifies the company's appeal to prospective buyers, protects its reputation, preserves revenue generation, and provides guidance during unplanned business interruptions. For buyers, confirming the BC/DR preparedness of the acquisition target safeguards the deal's integrity and ensures the long-term viability of the investment by protecting against future disruptions and operational instability.

Deal Impact: Buyers should be vigilant and assess whether there is a need for immediate or post-acquisition BC/DR enhancements. If BC/DR capabilities are lacking, adjusting the deal structure or valuation may be required to account for the cost, time, and effort required to address the issues.

Artifacts: Tangible documentation validating commitment to BC/DR should be obtained. Key artifacts may include the Business Impact Analysis report, which outlines the likely effects of disruptions and strategies in place to mitigate them, and a BC/DR Testing Results Document, which provides a detailed account of the testing scenarios, outcomes, and subsequent refinements made to the plan.

Summary

- Rationale: Robust BC/DR strategies ensure the organization's resilience amid disruptions and are foundational to operational stability.

- Insight: Events such as the 2017 WannaCry ransomware attack and the 2020/2021 pandemic have highlighted the importance of functioning BC/DR capabilities.

- Deal Impact: Weak BC/DR measures can threaten operational stability, which can impact the acquisition's valuation and feasibility.

Incident Management

Assessing an acquisition target's incident management plan is important for gauging its general preparedness and ensuring the deal's stability during the acquisition phase. An effective incident management plan provides a foundation for maintaining trust and transaction momentum if a cyber incident occurs during this phase. The widespread nature of cyberattacks and the implications of a target's ability to manage a cybersecurity incident during regular business operations are vital to its business health and the preservation of the deal.

Action: To prepare for M&A activities, sellers should have or establish a robust incident management plan and process. This involves crafting a systematic approach to detect, respond, and recover from security incidents, especially when managing incidents during the transaction period. Essential components include rapid detection capabilities, a dedicated incident response team, clear communication channels, and comprehensive post-incident analysis procedures. Training personnel and conducting regular drills, simulations, and incident response tests to validate the effectiveness of the plan are best practices for ensuring readiness.

Benefits: A well-structured incident management process offers strategic advantages for sellers, preserving the deal and the company's valuation if an incident occurs during the transaction phase or after closing. A robust incident response capability gives buyers confidence in the company's readiness to respond to security threats and breaches. This ensures that the company can minimize the repercussions of security incidents, particularly during the delicate transaction phase.

Security incidents are, by nature, disruptive. During the transaction, they can have an enormous impact. Buyers should review incident response plans and protocols and ensure that they are effective. Competent incident management practices are the only controls that can mitigate cybersecurity incidents if they occur during the transaction process.

Deal Impact: Buyers should conduct thorough due diligence on the target's incident management capabilities to mitigate the risks posed by cybersecurity incidents. If the incident response plan is inadequate, it may be necessary to factor in the costs of improving the plan through advanced tools, personnel training, or process design. Such costs might require a renegotiation of the acquisition terms or valuation.

Artifacts: Key artifacts should include the Incident Response Plan, which outlines the company's process for handling security incidents. Incident Logs and Post-Incident Analysis Reports provide transparency regarding the company's handling of past incidents and any subsequent improvements made to the plan. The results of documented incident response tests reveal the company's response capabilities under simulated conditions. These documents provide insight into a company's ability to manage security incidents, particularly during the transaction phase.

Summary

- Rationale: Prompt and effective responses to cyber incidents protect and preserve an organization's reputation, operational consistency, and financial health.
- Insight: The protracted response to the Equifax breach in 2017 worsened its impact, eventually costing the company $425 million in claims settlements.
- Deal Impact: Inefficient incident response can increase risks, leading to unforeseen post-acquisition challenges and costs. If an incident occurs during the deal, it can cause the deal to collapse.

Security Personnel

The cybersecurity team represents the company's dedication to addressing cyber threats and is the face of its cybersecurity strategy. In M&A situations, the expertise, structure, and efficiency of the cybersecurity team should not be overlooked. They affect both the immediate transaction and the cybersecurity posture of the merged or acquired entity. Both sellers and buyers should assess the skills and priorities of the cybersecurity staff to maintain cybersecurity resilience and maintain the strategic value of the deal after the transaction.

Action: Sellers should verify that their cybersecurity team has the right staffing, skills, and structure. They should offer consistent training and clearly define personnel roles for effective management of the company's security program. In contrast, buyers should examine the composition, expertise, and efficacy of the seller's cybersecurity team. They should understand the team's strategic direction and daily security procedures. This examination involves assessing the team's ability to manage threats, facilitate internal training, execute awareness campaigns, and the influence of these processes on the cybersecurity culture of the organization.

Benefits: A competent security team that consistently mitigates threats signals a mature approach to cybersecurity management to buyers. For sellers, this can enhance appeal and expedite the due diligence process by reducing uncertainties and complications posed by persistent threats. For buyers, evaluating the proficiency of the security team offers a glimpse into the company's cybersecurity maturity. This assessment helps determine if additional resources are needed after the transaction to ensure that cybersecurity measures align with business goals.

Deal Impact: Purchasers should consider the implications of the findings of the cybersecurity team assessment. If gaps in security staffing or expertise are identified, it could influence the overall value and structure of the deal. Buyers might suggest improvements to the security team pre-acquisition, adjust the final acquisition price to account for future enhancements to the security team, or propose holdbacks for future revenue impacts due to the deficiencies.

Artifacts: Key artifacts may include the Security Personnel Organizational Chart, Training and Certification Records, Job Descriptions, and Performance Reviews. These provide a comprehensive view of the team structure, learning efforts, expectations, and performance. These documents provide a transparent view of the investment in security personnel and their capabilities. They support validation of the seller's commitment to cybersecurity.

Summary

- Rationale: Investing in cybersecurity personnel reflects an organization's commitment to effectively managing cyber risks.
- Insight: Entities with well-staffed and skilled cybersecurity teams are generally less prone to damaging breaches.
- Deal Impact: Inadequate cybersecurity staffing or expertise can indicate an underprioritized security approach, influencing the value and structure of the deal.

Chapter 10

MAC-F: Manage

Security Documentation & Governance

Security documentation and governance are the foundation of an organization's cybersecurity framework. They reflect its commitment, strategy, and operational approach to managing cyber risk. Security documentation provides transparent and tangible insights into the cybersecurity maturity, culture, and strategy of the target company. The effectiveness of security governance is reflected in the guidelines, standards, processes, and policies that document how the organization's information assets are protected. It should reflect clearly defined roles and responsibilities for cybersecurity personnel and strategic alignment with the organization's IT and business strategy. Deficiencies in this domain are a latent risk to the transaction and will carry over to the merged or acquired entity if they are not mitigated.

Action: Sellers must consolidate, review, and refine all security documentation, ensuring that policies, procedures, standards, and guidelines are current, relevant, and aligned with IT and business objectives, industry best practices, and regulatory requirements. Governance processes must be evaluated and reinforced to ensure effective oversight, accountability, and continuous improvement. Buyers should review security documentation and evaluate the company's governance structure.

Benefits: For sellers, prioritizing security documentation and governance enhances attractiveness during M&A discussions and mitigates transaction risks by providing a structured and verifiable approach to managing cybersecurity. For buyers, a thorough evaluation of the seller's documentation and governance provides insight into the organization's cybersecurity maturity and any gaps that may exist.

Deal Impact: During deal negotiations, buyers should consider the implications of any identified gaps or weaknesses in a seller's security documentation and governance. This might involve factoring in remediation costs, such as investments in resources, training, or external consultants, into the final acquisition price or proposing adjustments to the deal structure to account for acquisition or post-merger risks.

Artifacts: Key artifacts might include security policies, Monitoring Policies, User Acceptance Policies, procedures, guidelines, and Governance Reports detailing oversight activities, audit findings, and continuous improvement initiatives. These artifacts provide a transparent and verifiable overview of the seller's cybersecurity governance and documentation and enables buyers to assess the cybersecurity posture of the entity.

Summary

- Rationale: Comprehensive security documentation and governance reflect the commitment of an organization and its strategic approach to cybersecurity management.
- Insight: Effective security governance ensures that cybersecurity strategies and operations remain relevant, is ready to respond to evolving threats, and consistently mitigates cyber risk.
- Deal Impact: Gaps or weaknesses in security documentation and governance can indicate vulnerabilities or a misalignment between stated policies and actual practices, influencing the value and structure of the deal.

Management and Board Focus on Cybersecurity

The dedication of an organization's leadership to cybersecurity is a testament to its core values and strategic priorities. In M&A situations, this commitment becomes essential. It emphasizes the organization's intent to safeguard digital assets, uphold its reputation, and maintain operational stability. Buyers must evaluate the participation of top leadership in cybersecurity. They should ensure that there is a clear cybersecurity focus, demonstrated by tangible oversight, and aligned with the organization's broader business objectives and risk management strategies.

Action: Sellers should ensure that cybersecurity and its strategic importance to the business are regularly discussed and highlighted in management and board meetings. Sellers should present updates, risks, and strategies to the leadership and board. They should also actively involve them in setting the cybersecurity direction, approving budgets, and aligning with broader business goals. Buyers should closely examine these practices to ensure that tangible actions, strategic decisions, and allocated resources validate the stated commitment.

Benefits: Active management and board participation in cybersecurity can make a company more attractive in M&A discussions. It shows a top-down strategic approach and alignment between cybersecurity strategies and business objectives.

Deal Impact: Buyers should evaluate the commitment of the board and executive management to cybersecurity during deal negotiations. Discrepancies between stated commitment and actual implementation may warrant changes to the deal's valuation or structure. These changes should take into account any identified risks or the need to improve cybersecurity strategy and operations.

Artifacts: Essential documents could include Board Meeting Minutes that detail cybersecurity discussions, decisions, and directives. These minutes provide evidence regarding security updates and strategies discussed among the company's leadership. In addition, training records for the board and management can demonstrate ongoing education and commitment to staying up to date with emerging cybersecurity threats and recommended practices.

Summary

- Rationale: A tangible commitment to cybersecurity at the management and board levels signifies strategic importance and organizational prioritization.

- Insight: The cascading effect of leadership commitment fosters a robust cybersecurity culture throughout an organization.
- Deal Impact: Discrepancies between the proclaimed commitment and actual practices can indicate vulnerabilities or misalignments, influencing the overall valuation and structure of the deal.

Organizational Awareness and Cybersecurity

Organizational awareness of cybersecurity addresses the human element and should be viewed as part of the company's defensive capabilities and its company-wide commitment to protect assets and data. Ensuring that every individual within the organization is aware of their role in maintaining cybersecurity and is knowledgeable in recognizing and mitigating threats is an effective defense against social engineering and insider threats. Sellers must evaluate and cultivate organizational awareness, ensuring that the company complies with cybersecurity best practices and is resilient to threats.

Action: Sellers preparing for a merger or sale should prioritize fostering robust cybersecurity awareness across the organization. This involves rolling out regular training sessions, workshops, and awareness campaigns for employees at all levels, ensuring that everyone knows their role in maintaining cybersecurity and is adept at recognizing threats. During their due diligence, buyers should validate the efficacy and penetration of these awareness programs, ensuring that the organizational culture is inherently aligned with cybersecurity best practices and that the workforce can recognize and report threats.

Benefits: For sellers, cultivating and demonstrating a high level of organizational cybersecurity awareness improves the company's position during M&A negotiations. This protects the company against threats and assures buyers that the human element of threat defense is not overlooked.

Deal Impact: Buyers should consider the depth and efficacy of the organizational awareness programs of the target company during deal negotiations. This could involve evaluating a company's historical response to threats, the penetration of its awareness programs, and the readiness for future threats. Any gaps or inefficiencies in the company's organizational awareness and cybersecurity training programs may require adjustments in the value or terms of the deal, ensuring that the buyer does not assume undue risk or inherit a weakened cybersecurity posture.

Artifacts: Training and workshop materials are essential documents that validate a company's efforts in enhancing organizational awareness. These materials shed light on the nature and regularity of cybersecurity awareness programs. Employee acknowledgment forms or certificates validate individual involvement and comprehension. The Insider Threat Program Documents detail the company's methods for detecting, reducing, and addressing insider threats. These records give buyers a clear perspective on the company's dedication to sustaining organizational awareness and addressing insider threats.

Summary

- Rationale: Fostering organizational awareness reduces the risk of breaches resulting from human error or insider threats and reflects a company-wide commitment to cybersecurity.

- Insight: A well-informed workforce acts as the first line of defense against various cybersecurity threats and can be particularly reassuring to buyers during M&A discussions.
- Deal Impact: Demonstrating a company-wide commitment to cybersecurity awareness can protect a company's assets, reputation, and valuation during M&A negotiations.

Chapter 11

MAC-F: Comply

Customer Contracts

Customer contracts reflect revenue and may include cybersecurity compliance requirements. Ensuring compliance with cybersecurity and data protection clauses within customer contracts is paramount to safeguarding customer relationships, maintaining regulatory compliance, and ensuring deal integrity. Sellers and buyers should thoroughly examine the cybersecurity terms in customer contracts. They should validate that the company is compliant with current and, is ready to fulfill, future customer security obligations.

Action: Sellers should thoroughly review customer contracts, focusing on cybersecurity and data protection terms, to confirm compliance. They should confirm that they are following all data management and cybersecurity provisions and are aware of noncompliance penalties. During due diligence, buyers should verify the seller's adherence to cybersecurity provisions in customer contracts.

They should ensure there are no hidden risks or instances of noncompliance that could negatively affect the deal's value or viability.

Benefits: Sellers should confirm the adherence to cybersecurity terms in customer contracts to enhance their position in M&A negotiations. Compliance with cybersecurity provisions in customer contracts indicates a commitment to customer satisfaction. It also protects the company from related financial penalties and reputational harm. Moreover, it ensures a seamless transaction by reducing the likelihood of related non-compliance issues impacting the merger or acquisition. Buyers should confirm that the entity is compliant with the cybersecurity provisions of its customers' contracts. Doing so reduces the risk of hidden customer-related cybersecurity issues that may undermine the deal and devalue the acquisition after closing.

Deal Impact: Buyers should consider the impact of any instances of noncompliance with cybersecurity provisions in customer contracts during deal negotiations. This could involve adjusting the valuation of the deal to account for identified risks, financial penalties, or the costs of the necessary improvements to cybersecurity measures to ensure future compliance.

Artifacts: Key artifacts might include a Customer Contract Compliance Report, which provides a detailed overview of the company's compliance with cybersecurity clauses within customer contracts. Additionally, any correspondence or documentation related to past noncompliance, remediation measures, and communication with customers regarding cybersecurity matters should be compiled and reviewed.

Summary

- Rationale: Ensuring compliance with cybersecurity clauses within customer contracts safeguards against legal, financial, and reputational risks, ensuring customer confidence and transaction integrity.
- Insight: Past instances of noncompliance or difficulties adhering to cybersecurity clauses can indicate future risks and should be thoroughly evaluated and addressed.
- Deal Impact: Noncompliance with cybersecurity clauses within customer contracts can jeopardize the value and feasibility of the deal, leading to financial penalties, loss of customer confidence, and reputational damage.

Audits and Regulatory Compliance

A company's track record of compliance with cybersecurity regulations and its audit performance record are key indicators of its cybersecurity posture, operational discipline, and risk management capabilities. Examination of audit and regulatory compliance records provides a snapshot of the company's current cybersecurity readiness and provides information on its cybersecurity and compliance program. Deficiencies in this domain can have a negative impact on the transaction, valuation, and the post-transaction entity.

Action: Sellers should meticulously review all past and current audits, regulatory penalties, and sanctions to ensure that all findings have been addressed and corrective actions have been implemented. This involves documenting the findings of both internal and external audits, the company's responses, and any subsequent actions taken to correct identified issues. In addition, any regulatory penalties, their causes, and the company's responses must be thoroughly documented and reviewed to ensure continued compliance and prevent future penalties.

Benefits: Transparently addressing and providing detailed documentation on past audits and regulatory penalties can improve credibility in M&A discussions. It demonstrates a commitment to compliance, continuous improvement, and maturity in managing cybersecurity and regulatory requirements. Furthermore, providing buyers with a clear and comprehensive view of past audits and regulatory interactions reduces uncertainties and complications during the sale. It provides transparency into the company's compliance posture and its ability to learn from past issues and improve its compliance program.

Deal Impact: Reviewing the target company's record with audits and regulatory compliance provides insights into its cybersecurity and compliance posture. Any gaps, issues, or historical noncompliance instances identified during due diligence may be factored into the deal negotiations. This could involve adjusting the valuation of the deal to account for the future risks or costs associated with ensuring ongoing compliance and addressing any identified issues.

Artifacts: Key artifacts should include Audit Reports, which detail the findings, recommendations, and company responses to audits; as well as Regulatory Penalty Documents, which provide insight into any penalties or sanctions faced, their causes, and corrective actions taken. Additionally, Compliance Certificates or Statements should be compiled and reviewed that attest to the company's current compliance status with relevant regulations and standards.

Summary

- Rationale: Audits and regulatory examinations provide an impartial and external view of a company's cybersecurity and compliance preparedness.
- Insight: The responses of a company to audit findings and regulatory penalties provide information on its commitment to continuous improvement and compliance.
- Deal Impact: Historical non-compliance issues identified in audits can affect the deal's valuation and terms and may indicate future risks and costs.

Vendor Management

Managing vendor-related cyber risk is often overlooked but it is a requirement for full coverage cybersecurity risk management. Integration with and dependence on vendors for services and products is a necessity for competitive business operations. However, they are also vectors for cybersecurity threats that may cause breaches in unsuspecting companies. Vendor-related threats can also negatively impact a deal during the transaction and post-transaction period.

Action: Sellers should thoroughly assess their vendor dependencies and identify and evaluate all third-party vendors, particularly those with access to data or IT infrastructure. A comprehensive risk assessment for each vendor, periodic reviews of their security programs, and adherence to company cybersecurity standards are imperative. Additionally, vendor agreements should be crafted to delineate security expectations, responsibilities, and breach notification protocols, ensuring that a company's cybersecurity posture is not compromised through vendor relationships.

Benefits: Effective vendor management can be a key value proposition for sellers in M&A discussions. Demonstrating a robust understanding and management of third-party risks assures buyers that the company is committed to maintaining a broad cybersecurity scope, extending to vendor relationships. This reduces the risks associated with vendor-related cybersecurity incidents and ensures that third-party relationships do not introduce unforeseen vulnerabilities or complications during the transaction phase.

Deal Impact: Buyers must understand the target company's vendor management program and associated cybersecurity considerations. A vendor management program that proactively addresses cybersecurity confirms a company's recognition of external risks. If the company does not have a vendor management program, buyers can seek adjustments to the offer price to account for cybersecurity risks and implementation costs.

Artifacts: To validate a company's vendor management program, key artifacts should include Vendor Risk Assessment Reports detailing the security posture and risks associated with each vendor. Vendor Agreements or Contracts should be reviewed to ensure that the security clauses and expectations are clearly outlined. Additionally, Vendor Security Review Logs, which provide a chronological account of the regular security reviews conducted on vendors, serve as documentation of a company's ongoing commitment to managing vendor dependencies securely.

Summary

- Rationale: The cybersecurity posture of third-party vendors can directly impact the cybersecurity health of an organization and introduce vulnerabilities.
- Insight: Historical cybersecurity incidents, such as the Target breach in 2013, underscore the importance of stringent vendor cybersecurity assessments.
- Deal Impact: Vendor dependencies and their associated cybersecurity considerations can influence the deal's valuation and terms and introduce unforeseen risks after the acquisition or merger.

Conclusion

Cybersecurity is an indispensable strategic priority during M&A processes, underpinning the integrity and value of the deal. Sellers must substantiate their cybersecurity posture, maturity, and resilience through extensive evaluations across security domains within the categories of Protect, Detect, Respond, Manage, and Comply with the MAC-F. Comprehensive cybersecurity due diligence offers insights into a company's readiness to safeguard against and respond to current and emerging threats.

Understanding an acquisition or merger target's cybersecurity capabilities enables buyers to make decisions aligned with risk tolerance. Buyers must thoroughly evaluate a target company's cybersecurity health to determine gaps that may undermine operations, compliance, customer relationships, and overall resilience post-transaction.

Scrutiny of audit findings and regulatory compliance provides further insights into positioning for future challenges.

Collaborative prioritization of cybersecurity is imperative throughout the process. Sellers and buyers should work together to ensure seamless continuity of business operations, manage third-party risks, and sustain stakeholder confidence during the transaction and transition. Validating cybersecurity clauses in customer contracts further safeguards relationships and revenue.

Vigilance from both parties reduces risk and supports the target company's stability and growth trajectory. Comprehensive cybersecurity preparation and due diligence are the foundation upon which the stakeholders can build to sustain the company's growth and resilience. For sellers, it preserves value. For buyers, it protects value. And for the company, it sustains value into the future.

Visit *https://www.mac-framework.com for more information on using the MAC-F for your organization.*

Chapter 12

MAC-F Due Diligence

A few years ago, I advised a client in a regulated industry whose customer contracts mandated adherence to the Payment Card Industry Data Security Standard (PCI DSS). These cybersecurity clauses require the infrastructure that hosts customer data to implement rigorous controls that meet PCI DSS requirements. This security standard requires the implementation of specific cybersecurity controls to protect credit cardholder data.

The service contracts with the company's customers contained clauses for performing security audits of my client. When an audit uncovered nonadherence to the data protection clauses stipulated in the contracts, customers were entitled to suspend service payments until my client achieved compliance—furthermore, the contracts allowed for termination of the agreement without penalties.

When I engaged with the organization, one of its customers had already indicated that they would not renew their contract due to security noncompliance, and another customer had stopped payments for seven months. The organization worked diligently to mitigate the issues and eventually became compliant. However, noncompliant cybersecurity issues cost the firm about 19 percent of its annual revenue for that fiscal year.

This episode would have had a much larger impact on the company if it had occurred during an M&A transaction. It underscores the importance of reviewing cybersecurity clauses or mandates in customer contracts, ensuring compliance, and avoiding negative revenue impact.

Despite the complexities of the M&A process, the underpinning motivation is that the deal represents significant value creation for buyers and sellers. A crucial period occurs after the parties sign the Letter of Intent. This initiates the due diligence phase, often constrained to tight timelines of 90 days or less.

Within this compressed timeframe, the buyer seeks assurance against any financial, legal, and operational pitfalls including those emerging from cybersecurity threats and vulnerabilities. The MAC-F, described in Chapter 5, manages cyber risk during this period. It is a pragmatic framework that addresses cybersecurity risks during and after an M&A transaction without overshadowing the ongoing security program of the target entity.

The objective of the MAC-F is to provide the buyer with timely information regarding the cybersecurity posture of the target merger or acquisition. It enables buyers to make informed decisions regarding identified cybersecurity risks and their impact on deal valuation, viability, and post-transaction value. Sellers can confidently prepare for an exit or merger by applying the MAC-F to mitigate cybersecurity risk and demonstrate cyber readiness to buyers. The following case study illustrates how the MAC-F is applied to manage cyber risk in M&A.

The case study centers on the execution of the MAC-F by the buy side during due diligence. The sell side executes the MAC-F before engaging buyers or due diligence in preparation for a sale or merger.

This case study examines the acquisition of Medical Technology Data Systems (Medical Tech) by Millennium Private Equity (Millennium) using the MAC-F to perform due diligence and obtain transparency regarding cyber risk and fiscal impact.

MAC-F Due Diligence Case Study: Millennium Private Equity's Acquisition of Medical Tech Data Systems

Millennium Private Equity (Millennium) has grown its impressive portfolio of companies in the biotechnology sector by making strategic acquisitions and has a strong financial performance record. Recognizing cybersecurity as a foundation for safeguarding intellectual property, customer data, and operational integrity, Millennium ensures that acquired entities demonstrate cyber resilience and financial viability.

Medical Tech Data Systems (Medical Tech) is an attractive target, with $100 million in annual revenues and robust healthcare data management capabilities. Serving research institutions, healthcare service providers, and pharmaceutical companies, Medical Tech houses a wealth of sensitive data, making cybersecurity a top priority.

The offer price of ten times revenue underscores Medical Tech's value proposition.

There is a clear strategic alignment between the companies. Millennium recognizes Medical Tech not only as a profitable investment, but also for its potential to unlock synergies across the biotechnology portfolio through enhanced data-driven decision-making. The Medical Tech founders will enjoy a rewarding exit after years of hard work building the company. However, Medical Tech's vast proprietary datasets are highly coveted by adversaries and require stringent protection. Therefore, the acquisition is required thorough cyber-diligence.

This case study illustrates how Millennium leveraged the MAC-F to conduct comprehensive cybersecurity due diligence before the acquisition. This ensured cybersecurity risk transparency and risk mitigation during and after the transaction.

Phase 1: The MAC-F Due Diligence Assessment

Using the MAC-F, an assessment is conducted to evaluate the cyber-security acquisition preparedness of Medical Tech.

Step 1: Evidence Collection

Executing due diligence involves a thorough assessment and collection of evidence across each security domain, encompassing the following:

- Policies and procedures: Review all relevant documentation related to security policies, standards, guidelines, and procedures.

- System Configurations: An assessment of technical controls and system configurations.
- Interviews: Engagements with key personnel to obtain information about existing security practices.
- Logs: Examine security logs, audit trails, and other historical security data.
- Additional Information: Exploration of any other relevant artifacts or evidence that may reveal the security posture for each domain.

Step 2: Risk Identification and Categorization

Identify and categorize cybersecurity risks according to the MAC-F categories. This classification provides a structured, high-level overview and facilitates a systematic approach to subsequent risk management activities. Risk is qualitatively rated based on its impact on the deal and post-closing if the risk materializes: High (major impact), Medium (moderate impact), and Low (minor impact). Risk is identified and assessed in specific security domains, such as Asset Management and Identity Management, and categorized within the MAC-F categories of Prevent, Detect, Respond, Manage, and Comply.

Step 3: Assessing Impact

The impact of each identified security risk is assessed in terms of the pending M&A deal and post-transaction implementation.

- Deal Impact: Considers how vulnerabilities might affect the deal's valuation, due diligence, negotiations, or legal aspects.

- Post Impact: Analysis of the prospective impact of security risk after the deal closes and integration or handover begins.

Step 4: Mitigation

Mitigation strategies were formulated for each identified security gap and weakness, and the associated costs were estimated.

- Mitigation: Development of strategies to address identified security gaps and weaknesses.

Step 5: Cost Estimation

- Capital Cost: Estimate the one-time costs required to implement the recommended risk mitigation activities.
- Recurring Cost: Estimate ongoing annual costs to sustain mitigation measures.

The matrix that follows illustrates the data collected for phase 1 of Millennium's MAC-F due diligence of Medical Tech.

Category	Security Domain	Diligence Evidence	Deal Impact	Post Impact	Mitigation	Capital Cost	Recurring Cost
Prevent	Asset Management	No Asset Inventory or Policies in place (Policies & Procedures)	High - Inability to accurately value assets	High - Asset loss/theft, audit failures	Develop asset management program and inventory management program	$75,000	$25,000
	Identity Management	MFA is not enabled for sensitive systems (Systems Configurations)	Medium - Account takeover probability before the merger is complete	Low - Can enable MFA post-acquisition	Roll out MFA tokens to all admin accounts	$25,000	-
	Data Protection	Unencrypted sensitive data stored (Logs)	High - Data breach could derail the deal	High - Ongoing risk of data exfiltration	Implement data encryption and DLP controls	$250,000	$75,000
	Network Security	Minimal Network Segmentation (System Configurations)	Medium - Risk of lateral movement	Medium - Ongoing access risk if not segmented	Redesign of network architecture with segmentation	$100,000	$25,000

Prevent	Systems Security	Weak System Password Policies (Policies & Procedures)	Medium - Potential for unauthorized access	Low - Can implement strong controls post-merger	Enforce stringent password policy across systems	-	-
	Endpoint Security	No endpoint detection controls (Interviews)	High - Inability to detect threats	High - Persistent endpoint vulnerabilities	Deploy EDR tools across endpoints	$150,000	$50,000
	Application Security	Sensitive applications lack input validation (Interviews)	Medium - Data integrity issues possible	Medium - Ongoing risk of exploitation	Conduct code review to remediate flaws	$200,000	-
	Physical Security	No visitor management system (Site Visit)	Low - Can implement post-acquisition	Medium - Ongoing access risks	Install visitor management system	$50,000	-
Detect	Monitoring	No centralized log management (Tools Evaluation)	Medium - Limited visibility into threats	High - Lack of threat visibility if not addressed	Implement SIEM platform with log collection	$200,000	$50,000
	Vulnerability Mgmt.	No vulnerability scanning performed (Interviews)	Medium - Vulnerabilities may impact asset valuation	Medium - Ongoing cyber exposure if not addressed	Implement monthly vulnerability scanning	$25,000	$5,000

Detect	Past Incidents	No documentation or analysis of previous incidents (Interviews)	Low - Can conduct analysis post-acquisition	Medium - Lack of historical learning	Perform incident analysis on previous breaches	$50,000	-
	Business Continuity	No BC/DR plans or testing (Interviews)	Low - Can implement post-acquisition	High - Lack of resilience to disruptions	Develop business continuity plans and test them annually	$100,000	-
Respond	Incident Response	No documented IR plan or playbooks (Documentation Review)	High -Develop an IR plan after the acquisition	High - Lack of readiness will delay effective response	Develop IR plan with playbooks, test annually	$75,000	$10,000
	Security Personnel	No dedicated cybersecurity resources (Interviews)	Medium - Leadership gaps may impact the diligence process	High - Lack of skilled resources to manage risk	Hire CISO and key cybersecurity roles	$200,000	$500,000
Manage	Security Governance	No security policies established (Documentation Review)	Medium - Lack of guidance on security practices	Medium - Ongoing inconsistencies without policies	Define information security policies	$75,000	-

Manage	Management Focus	No security metrics provided to executives (Interviews)	Low - Can brief executives post-acquisition	Medium - Leadership lacks visibility without metrics	Develop an infosec metrics program tied to risk	$100,000	$25,000
	Security Awareness	No security awareness program (Interviews)	Low - Training can be implemented post-merger	Medium - Workforce vulnerabilities without training	Develop a security awareness program	$50,000	$25,000
Comply	Customer Contracts	No review of customer contracts and obligations (Contracts Review)	Medium - Potential for non-compliance	Medium - Ongoing compliance gaps	Review contracts and evaluate security obligations	$25,000	-
	Industry Regulations	No audit or gap assessment was performed (Interviews)	Medium - Potential for non-compliance	Medium - Ongoing non-compliance risk	Conduct an audit of regulatory compliance	$75,000	-
	Vendor Management	No vendor security assessments (Interviews)	Medium - Third-party risk may impact diligence	Medium - Supply chain disruption if vulnerabilities are exploited	Initiate vendor risk assessment program	$75,000	$25,000

Phase 2: The MAC-F Risk Analysis Summary

The following summary provides a detailed financial and strategic overview of cybersecurity considerations in the acquisition of Medical Tech by Millennium. This section presents analyses within the cybersecurity categories of Prevent, Detect, Respond, Manage, and Comply. It highlights the high-risk areas, analyses pertinent to the deal, and post-acquisition operations, as well as associated capital and recurring costs. The objective is to ensure that the acquisition is secure from the identified cyber threats, vulnerabilities, and associated financial pitfalls.

Prevent

- High-Risk Domains: Asset Management, Data Protection, Endpoint Security
- Analysis: Major risks arise from the lack of a robust asset management program, sensitive unencrypted data, and the absence of endpoint detection controls, which could impact the value and synergy of the deal.
- Mitigation Summary:
    - Asset Management: Establish a comprehensive asset management program.
    - Data Protection: Implement data encryption and Data Loss Prevention (DLP) controls.
    - Endpoint Security: Deploy Endpoint Detection and Response (EDR) tools.
- Capital Cost: $850,000
- Recurring Cost: $175,000

Detect

- High-Risk Domains: Monitoring
- Analysis: The absence of centralized log management and vulnerability scanning mechanisms could allow threats to remain undetected, posing a latent risk to the transaction.
- Mitigation Summary:
    - Monitoring: Implement a Security Information and Event Management (SIEM) platform and conduct regular vulnerability scans.
- Capital Cost: $275,000
- Recurring Cost: $55,000

Respond

- High-Risk Domains: Security Personnel, Incident Response
- Analysis: The lack of dedicated cybersecurity resources and a documented incident response (IR) plan could delay effective responses to incidents, escalating isolated incidents into organization-wide crises.
- Mitigation Summary:
    - Security Personnel: Hire key cybersecurity roles.
    - Incident Response: Develop an IR plan.
- Capital Cost: $375,000
- Recurring Cost: $510,000

Manage

- High-Risk Domains: Security Governance
- Analysis: The absence of established security policies and executive oversight metrics could lead to inconsistencies and a lack of guidance on security practices after the transaction.
- Mitigation Summary:
  - Security Governance: Define information security policies and develop an information security metrics program.
- Capital Cost: $225,000
- Recurring Cost: $50,000

Comply

- High-Risk Domains: Customer Contracts, Industry Regulations
- Analysis: Without a thorough review of customer contracts and an audit of regulatory compliance, transactions face noncompliance risks.
- Mitigation Summary:
  - Customer Contracts: Review contracts and evaluate security obligations.
  - Industry Regulations: Conduct an audit of regulatory compliance.
- Capital Cost: $175,000
- Recurring Cost: $25,000

Heatmap (Risk Level: Low, Medium, High)

Category	Security Domain	Deal Impact	Post Impact	Capital Cost	Recurring Cost
Prevent	Asset Management	High	High	Substantial	Moderate
	Data Protection	High	High	Substantial	Substantial
Detect	Monitoring	Medium	High	Substantial	Substantial
Respond	Security Personnel	Medium	High	Substantial	Substantial
	Incident Response	High	High	Moderate	Minimal
Manage	Security Governance	Medium	Medium	Moderate	Moderate
Comply	Customer Contracts	Medium	Medium	Moderate	Minimal

- **Minimal:** < \$50,000
- **Moderate:** \$50,000 - \$150,000
- **Substantial:** > \$150,000

The analysis underscores the criticality of addressing the identified risks, particularly in the "Prevent" and "Respond" categories, to safeguard the deal from disruptions and ensure the security and compliance of Medical Tech Data Systems post-transaction.

Prompt attention to high risk domains is recommended—particularly Asset Management, Data Protection, Security Personnel, and Incident Management—to mitigate the dangers affecting deal valuation and post-transaction operations.

Phase 3: The MAC-F Financial Analysis Summary

Millennium Private Equity understands that the capital and recurring costs associated with mitigating cybersecurity risks should be factored into deal negotiations and post-transaction budgeting to ensure the financial viability and success of the acquisition. The transparency regarding cybersecurity risks provided by MAC-F allows Millennium to model three different scenarios for its valuation adjustment. The details of the analysis are as follows.

Overview of Capital and Recurring Costs

Capital Costs

- Total Capital Costs: $1,900,000
- Breakdown:
  - Prevent: $850,000
  - Detect: $275,000
  - Respond: $375,000
  - Manage: $225,000
  - Comply: $175,000

Recurring Costs

- Total Recurring Costs: $815,000
- Breakdown:
  - Prevent: $175,000
  - Detect: $55,000
  - Respond: $510,000
  - Manage: $50,000
  - Comply: $25,000

Financial Impact on Deal Valuation

Initial Acquisition Price: $1 billion (10 times the revenue of $100 million)

Adjusted Acquisition Price Proposal: Given the substantial capital and recurring costs identified through the MAC-F framework, Millennium Private Equity is advised to negotiate the acquisition price to account for these imminent expenses and associated risks.

Capital Cost Impact: $1.9 million is required for immediate mitigation strategies. This is a direct upfront cost that will impact financials immediately upon acquisition.

Recurring Cost Impact: $815,000 annually to sustain mitigation measures. This is an ongoing cost and will impact annual revenues, thus affecting profitability and ROI.

Valuation Adjustment Proposals

- Proposal 1: Deduct the total capital and recurring costs for the first year from the acquisition price:
  New Price = $1 billion - ($1.9 million + $815,000) = $997.285 million.

- Proposal 2: Adjust the valuation multiplier considering the impact of recurring costs on annual revenue:
  New Annual Revenue = $100 million - $815,000 = $99.185 million.
  New Price = 10 * $99.185 million = $991.85 million.

- Proposal 3: A hybrid approach in which the capital costs are deducted from the acquisition price and the valuation multiplier is adjusted for the reduced revenue due to recurring costs:
  New Annual Revenue = $100 million - $815,000 = $99.185 million.
  New Price = (10 * $99.185 million) - $1.9 million = $989.95 million.

These proposals aim to adjust the acquisition price to more accurately reflect the financial impact of identified cybersecurity risks and associated mitigation costs. The negotiation should be transparent, highlighting the identified risks and the mitigation required to facilitate mutually beneficial agreements between both entities.

The financial implications of the cybersecurity risks identified through the MAC-F framework are substantial and warrant careful consideration in deal negotiations. Capital costs necessitate immediate expenditure, while recurring costs persistently impact revenue and, consequently, the valuation of Medical Tech.

The proposed adjustments to the acquisition price aim to account for these financial impacts, ensuring that Millenium does not overcapitalize and can secure a deal that safeguards its financial interests while enabling the strategic acquisition of Medical Tech. The parties agree that the negotiation should be transparent, highlighting the identified risks and the mitigations required to facilitate a mutually beneficial agreement.

Cybersecurity Financial Impact and Strategic Negotiation in M&A: A Case Study Insight (Definitive Agreement)

Based on the MAC-F Risk and Financial Analysis, Millennium Private Equity revised its valuation offer for Medical Tech to account for the costs associated with the cybersecurity risks identified during due diligence.

Initial Valuation, Capital, and Recurring Cybersecurity Costs

Millennium initially pegged the valuation of Medical Tech at $1 billion, considering its $100 million revenue and applying a 10x valuation multiplier. However, the MAC-F cybersecurity assessment revealed a capital cost of $1.9 million and an additional recurring cost of $815,000 per year, which would affect revenues and valuation.

Re-evaluation and Negotiation Proposals

Based on the Mergers and Acquisitions Cybersecurity Framework due diligence risk and financial reports, Millennium and Medical Tech have agreed to recalibrate and adjust for cybersecurity costs as follows:

- Adjusted Revenue: $100 million - $815,000 (recurring cost) = $99.185 million.
- Revised Valuation: $99.185 million * 10 = $991.85 million

Moreover, $1.9 million in capital costs for cybersecurity implementation further impacts the financial dynamics of the acquisition. Considering cybersecurity risks and financial impacts, Millennium, in its lead-up to signing the Definitive Agreement, has outlined the following options for Medical Tech that would enable mitigation of the cybersecurity risks identified during due diligence.

Proposals to Adjust Acquisition Price

- Direct Price Adjustment: Propose a new acquisition price of $991.85 million, ensuring that the price reflects the net revenue and accounts for capital cybersecurity costs.

- Capital Cost Offset: Advocates to directly reduce the $1.9 million capital cost from the acquisition price, ensuring that Millennium is not burdened with immediate cybersecurity investments.

- Staggered Price Reduction: Suggest that a portion of the price be held in escrow, to be released after successful cybersecurity implementations and operations over a defined period.

- Cybersecurity Escrow Account: Establish an escrow account, funded by the seller, dedicated to covering cybersecurity capital and recurring costs for a specified period post-acquisition.

- Revenue Target Earn-out: Implement an earn-out structure, making a portion of the acquisition price contingent on achieving adjusted revenue targets, considering cybersecurity costs.

- Cybersecurity Milestone Incentives: Establish a deal where a portion of the acquisition price is contingent on achieving specific cybersecurity milestones within a defined timeframe after acquisition.

- Seller Financing for Cybersecurity Costs: The seller pays for the cybersecurity costs for a defined period after the acquisition, mitigating the direct impact on revenue and profitability.

These proposals protect Millennium from overpaying and ensure that the acquisition price accurately mirrors the net revenue and associated risks of Medical Tech. By including cybersecurity costs in the negotiation and valuation, the buyer assumes the initiative in controlling cybersecurity risk, and the seller demonstrates a tangible commitment to the transaction and post-transaction health of the enterprise.

This case study serves as a blueprint, demonstrating that a structured approach to understanding and negotiating cybersecurity costs can influence an acquisition strategy, ensuring that the investment is secure and financially sound.

For sellers, the case study provides a blueprint for applying the MAC-F to ensure cyber readiness and resiliency in preparation for the sale or merger of the company.

Post-Due Diligence Considerations

Cybersecurity in Definitive Agreement Negotiations

- Cybersecurity Clauses: The Definitive Agreement outlines the terms and conditions of a transaction; it is essential to include clauses that address cybersecurity risks and liabilities. This can range from stipulations on post-acquisition breach disclosures to responsibilities related to cybersecurity compliance.

- Valuation Adjustments: Issues identified during cybersecurity due diligence can influence the final valuation of the deal. Both parties must understand and negotiate these adjustments to ensure that they reflect the actual cybersecurity risks and costs.

Navigating Regulatory Approvals and Compliance

- Understanding Regional and Industry Regulations: Different regions and industries have specific cybersecurity regulations. Ensuring that the target company complies with these standards minimizes legal complications and fines post-acquisition.

- Data Protection: Business risk considerations and regulatory requirements make data protection a critical security priority. Improper data protection measures can result in an array of negative repercussions that include financial, reputational and legal impacts.

The Closing Process and Cybersecurity Handover

- Secure Transition: As assets, data, and systems are handed over, it is essential to ensure their secure transition. This includes encrypted data transfers, secure access handoffs, and safe decommissioning of redundant systems.
- Collaboration: The end of the M&A process is the beginning of the post-acquisition journey. In the case of mergers, a collaborative approach to cybersecurity, in which both parties actively share insights, resources, and best practices, can set the stage for a secure and productive future.

Buyers and sellers can effectively evaluate a company's cybersecurity preparedness using the MAC-F. This ensures transparent, cyber-secure, successful M&As that preserve strategic and economic value.

Visit *https://www.mac-framework.com for more information on using the MAC-F for your organization.*

Chapter 13

Securing the Transition

The city of Boston, with its rich history and vibrant tech scene, was the backdrop for a meeting that would shape the future of a midsize technology firm. As my plane touched down that Sunday evening, I was acutely aware of the challenges ahead. The next day, I was scheduled to meet with the executive team and the newly appointed CISO of a company that had recently undergone substantial changes.

The firm, whose core business was to analyze billions of credit card transactions, had recently been acquired by a larger corporate entity. Although promising on several fronts, this acquisition led to unexpected complications. Many members of the technology and security teams, including the CIO and the previous CISO, had chosen to leave. Their departure created a vacuum, leaving the company's cybersecurity posture uncertain and vulnerable.

Upon entering the meeting room that Monday morning, I could sense the weight of the situation. The faces of those around the table, including the new CISO, were filled with concern, uncertainty, and despair. As I tried to gauge the cause of their anxiety, it became clear that an upcoming PCI DSS audit was the main source of their unease. This audit, scheduled in just four months, is a regulatory requirement for any company that manages credit card information. For this company, whose business focused on credit card data analysis, the audit was more than just a requirement—it was the foundation of their business model.

The stakes were incredibly high. A failed audit would have cascading repercussions. Their clients, a sizable percentage of which are leaders in their respective industries, could demand the cessation of data processing. The PCI Council, which is responsible for the standard, could impose substantive fines. But beyond these tangible setbacks, there was an even more far-reaching risk: the erosion of the company's hard-earned reputation. Any hint of cybersecurity lapses could be catastrophic in an industry where trust is central to business health.

Given the urgency, our first order of business was to understand the current state of their cybersecurity measures. We decided on a gap assessment, a process that would help identify areas of noncompliance or vulnerability. With the clock ticking, the assessment had to be swift yet thorough. The remaining technology teams, who were still grappling with the changes post-acquisition, were forthcoming. They verbally confirmed the existing security measures, highlighting strengths and areas of concern.

To ensure this transparency, the executive team demonstrated real leadership. They set the tone, assuring the technology teams that the goal was not to assign blame but to understand and rectify any existing issues. There would be no repercussions for honest disclosures, even if they revealed shortcomings. This assurance was liberating. It created an environment of trust where teams could openly discuss vulnerabilities without fear of reprisal.

With the gaps identified, the next phase was corrective action. The company mobilized internal and external resources to address the vulnerabilities. They revised protocols, introduced new security measures, and conducted intensive training sessions. The goal was clear: to ensure that by the time of the PCI DSS assessment, the company would meet and exceed the required standards.

Months of hard work and collaboration culminated in a momentous achievement: The company passed the PCI DSS assessment without any findings. It was a testament to their resilience, adaptability, and unwavering commitment to cybersecurity.

However, beyond this immediate success, the experience offered deeper insights. Mergers and acquisitions, while offering growth and new opportunities, can sometimes lead to unexpected challenges. In the euphoria of expansion and the intricacies of integration, enterprise functions like cybersecurity can sometimes be overshadowed. This incident was a stark reminder of the importance of maintaining a cybersecurity focus, regardless of organizational changes.

In today's data-driven age, cybersecurity vigilance is imperative for companies amid regular operations and corporate changes like mergers and acquisitions. With data breaches frequently making headlines, consumer trust is fragile. Cybersecurity is more than just data protection—it safeguards corporate reputation, customer confidence, and the very essence of the business itself. Companies must prioritize cybersecurity to demonstrate their commitment to customers and stakeholders.

In conclusion, the Boston meeting and the subsequent months were a journey of discovery, collaboration, and eventual triumph. It underscored the importance of effective cybersecurity measures, the value of transparency, and the need for adaptability in the face of challenges.

Mergers and acquisitions are transformative events in the business life cycle. They represent growth, change, and new beginnings. However, as with any notable change, they come with their own set of challenges. The handover phase is one of the most important junctures in this process.

Similar to the baton pass in a relay race, this transition must be controlled. Cybersecurity continues to be a focal point of this phase. As assets, data, and operations transition from seller to buyer, maintaining operational integrity is a top priority. This chapter focuses on the importance of cybersecurity during handovers.

Collaboration

Collaboration is the cornerstone of success. As companies merge or change hands, the transition is not only about assets and operations but also about trust, understanding, and shared objectives.

Transparent Communication

Open and transparent communication is the foundation for successful collaboration. Sellers should prioritize keeping the buyer informed of material changes in cybersecurity posture. This includes vulnerabilities, incidents, and steps taken to address them. In doing so, sellers build trust and ensure that there are no unpleasant surprises. This approach can minimize disputes and misunderstandings that could jeopardize the deal.

Joint Workshops

Bringing teams from both sides together to work through challenges can be beneficial. Organizing workshops where they can discuss, understand, and even challenge the existing security framework, tools, and protocols can lead to better post-acquisition integration. Such sessions also allow teams to align their cybersecurity objectives and strategies, ensuring a smooth transition and ongoing cooperation.

Documentation Handover

Documentation provides tangible evidence of a company's cybersecurity efforts. Providing complete records of all initiatives and programs, incidents, and assessments offers the buyer a clear picture of a company's cybersecurity posture. It also serves as a reference point for future cybersecurity initiatives to ensure continuity and consistency.

Business Relevance

A smooth transition is not just about transferring assets; it is also about building a foundation of trust. By collaborating closely with the buyer, sellers can ensure that they are well-equipped to sustain and enhance post-acquisition security. This reduces the likelihood of post-transaction disagreements and sets the stage for the future success of the combined or acquired entity.

Data Migration

Data is often the lifeblood of businesses. With the ongoing digital transformation of businesses, the importance of data cannot be overstated. As such, any data migration during M&A activities must be managed securely.

Data Audits

Before embarking on data migration, it is essential to plan the data transfer process. Comprehensive assessment and data classification can help identify the datasets that need to be moved and their sensitivity levels. This step ensures that only the necessary data is transferred and reduces risk. It also provides an opportunity to archive redundant or outdated data.

Secure Migration

Data breaches have catastrophic financial and reputational consequences. Using encrypted channels for data transfer is non-negotiable. This ensures that the data remains confidential and retains its integrity during migration. Employing secure migration channels also demonstrates the seller's commitment to data security to the buyer.

Validation

Once the data has been transferred, its integrity must be validated. This step ensures that no data is altered, corrupted, or lost during migration. This assures both the seller and the buyer that the value of the data remains intact.

Business Relevance

For buyers, the acquired data often constitutes a major part of the acquisition's value. They must ensure a secure and seamless migration to protect this value and reduce the risks of data breaches that can result in financial losses and reputational harm.

Holdback Clauses

Holdback clauses represent a portion of the purchase price that is "held back" or retained by the buyer for a specified period post-closing.

These clauses are designed to protect the buyer against unforeseen liabilities or ensure that certain conditions or representations made by the seller are met post-closing.

Holdback Conditions

The sell side should thoroughly understand the conditions under which the holdback amount is released. They should be clearly defined and actionable if tied to cybersecurity measures. This clarity allows sellers to prioritize and address them systematically, ensuring timely release of the held-back amount.

Cybersecurity Concerns

Sellers should take steps to mitigate identified issues if holdback clauses are tied to cybersecurity vulnerabilities or past incidents. This could involve patching identified vulnerabilities, improving the security infrastructure, or conducting third-party cybersecurity audits to validate security controls improvement or implementation.

Third-Party Audits

Consider initiating third-party cybersecurity audits to verify the implemented measures. This offers an impartial evaluation of the cybersecurity posture.

Security Personnel

During M&A, security personnel are often overlooked but play a key role in the success of the transition. These individuals are deeply embedded in the company and work diligently behind the scenes to protect assets, data, and operations during the transfer from seller to buyer. Their dedication ensures that the integrity and security of these elements remain uncompromised throughout the process.

The Role of Security Personnel

Security workers, particularly those who have been with the company for a long time, are more than staff members. They have a wealth of institutional knowledge, especially regarding a company's cybersecurity program. In some instances they have been present during the creation of security defenses, have first-hand experiences with past threats, and have been instrumental in crafting responses to them. Their positions allow them to establish relationships with various departments, vendors, regulatory bodies, and third-party service providers. These relationships, built on trust and collaboration, are essential during M&A transitions.

When a company undergoes mergers or acquisitions, it is necessary to maintain cybersecurity measures, protocols, and tools. Security personnel play a vital role in ensuring that they are intact, functional, and effective. As the Respond domain of the MAC-F demonstrates, security personnel have intimate knowledge of the technology environment and can address vulnerabilities to prevent exposure to threats.

Active Engagement

To ensure a smooth post-acquisition transition, it is advisable for the buyer to actively engage with the security team. This involves more than just a brief introduction; it requires a deeper understanding of their roles, responsibilities, and valuable insights. When security personnel are integrated into a new organizational structure, lasting benefits can be achieved. In the context of a merger, the latent knowledge of security personnel is essential to aligning the existing cybersecurity programs of both entities, whose objective is a cohesive and strong security culture.

Growth Opportunities

Because the cybersecurity landscape is constantly changing, it is wise to provide security teams with opportunities to improve their skills and knowledge. This includes advanced training programs, certifications, and exposure to the latest security tools and technologies. By investing in their growth and development, organizations can retain talent and ensure that their teams are prepared to face current cybersecurity threats.

Open Communication

A transition can be a period of uncertainty for employees, including security personnel. Fostering an environment in which they feel valued and heard boosts their commitment and performance. Establishing open communication channels where they can voice their concerns, provide feedback, or suggest improvements can be invaluable. Their extensive knowledge and practical experience may offer valuable insights that might be overlooked by the upper management or the buyer's team.

Continuity

During M&A, the focus is often on tangible assets, such as data and operations, while the importance of intangible factors, such as the expertise of security personnel, is often overlooked. However, ensuring a smooth transition for these professionals is highly recommended for effective human resource management and for safeguarding a company's cybersecurity posture. The seamless transition of security personnel is a wise move that protects the core of a company's security function. It ensures continuity in cybersecurity operations and minimizes disruptions, unaddressed threats and vulnerabilities.

Conclusion

The M&A handover phase bridges the seller's legacy with the buyer's future vision. Security employees, with their deep institutional knowledge, have navigated past threats and developed security strategies. Their relationships with departments, vendors, and regulatory bodies are invaluable. Ensuring cybersecurity continuity during this transition is paramount to safeguarding data, fostering trust, and laying a foundation for the combined entity's success. Emphasizing collaboration and secure asset transfer optimizes the outcome for both sellers and buyers.

Chapter 14

Artificial Intelligence – The Future is Now

Businesses are rapidly changing and adapting to technological advances and artificial intelligence (AI) is at the forefront of this revolution. From manufacturing to finance, the integration of AI is transforming industries in ways that we never thought possible. However, with these advancements, organizations face new threats, vulnerabilities, and risks.

The business of mergers and acquisitions is not immune to these changes. As AI-driven tools and solutions become more integral to M&A processes, understanding the threats posed by AI and the vulnerabilities of new AI tools will be a challenge.

The AI-Driven Business

AI has transformative powers that cannot be denied. It goes beyond automation and improves capabilities, strategies, and decision making. AI-driven tools have revolutionized traditional processes. Due diligence, which is time-consuming and laborious, is becoming faster and more accurate with AI-driven data analytics. Valuation models also benefit from AI's predictive capabilities, providing more precise and dynamic valuations based on vast datasets.

However, as the M&A sector becomes more dependent on AI, it will become more vulnerable to the unique threats that AI will introduce or amplify. Therefore, it is essential to understand and mitigate the associated risks to ensure the safe and effective integration of AI into M&A processes.

AI's Impact on M&A

AI has evolved from a futuristic idea into a tangible asset that businesses in various sectors can rapidly leverage. Its transformative power is also evident in M&A. The influence of AI on M&A processes is growing rapidly, from deal sourcing to post-merger integration. The shortlist of areas transformed by AI is described below.

Deal Sourcing

In the past, the identification of acquisition targets relied heavily on human networks and manual research, which is a time-consuming process that could lead to overlooking worthy targets. However, this process can become more efficient with the help of AI algorithms. AI can scan and analyze vast datasets, uncovering patterns, trends, and synergies that may go unnoticed in manual analyses. Consequently, AI streamlines the sourcing process and ensures that opportunities align with strategic goals.

Due Diligence

A thorough due diligence phase is an essential risk management activity in the M&A process. Before consuming the deal, buyers thoroughly examine the financial standing, operational efficiency, legal compliance, cybersecurity readiness, and other aspects of the target company. The ability of AI to process large data sets, perform analysis, and answer natural language queries using large language models (LLM) enables efficient and accurate identification of red flags. With AI, the intensive labor of the due diligence period is reduced, and a comprehensive understanding of inherent risks may be obtained.

Post-merger Integration

After a merger, several challenges must be addressed, such as aligning corporate culture and integrating IT systems. Predicting integration challenges can be difficult, but with AI's analytical and predictive capabilities, companies can devise more informed and proactive integration strategies. A predictive approach can minimize disruptions and the merged company can start providing value more quickly.

Embedding AI

Enhanced Due Diligence

As companies increasingly rely on AI assets, such as datasets, algorithms, and AI-driven processes, the scope of due diligence in M&A transactions is expanding. In addition to assessing economic value, M&A stakeholders must assess their AI security risks and compliance with evolving AI regulations. This introduces a new layer of complexity to the M&A valuation process, especially in the technology sector. AI assets can boost a company's valuation or lead to lower valuation adjustments if the AI assets are compromised.

Integration Risks

After the acquisition is completed, the integration of the AI-driven assets of the target company can introduce new vulnerabilities into the infrastructure of the acquiring company. These vulnerabilities can be due to outdated AI models, compatibility issues, and even inherent biases in algorithms. Addressing these risks requires a deep understanding of AI technologies and their effects.

Regulations

Regulators worldwide are formulating new regulations in response to AI's exponential growth and adoption. This evolving regulatory landscape introduces another layer of complexity to M&A transactions. Buyers and sellers of companies must stay abreast of the ever changing regulatory regime being introduced to address the challenges posed by AI.

AI Threats

Data Poisoning

Data accuracy is paramount for ensuring the reliability of AI. However, it is unfortunate that certain individuals may attempt to manipulate data, which can result in a phenomenon known as data poisoning. This occurs when subtle and misleading information is introduced, which adversely affects AI-driven predictions or analyses. In M&A transactions, the use of incorrect data can have dire consequences, such as the production of inaccurate valuation models. This can lead to poor decision-making and negatively affect the entire process. Therefore, it is necessary to prioritize the quality and integrity of data to avoid disastrous outcomes.

Model Inversion Attacks

Deep learning AI models are often viewed as opaque because they process input data and output results without clear intermediate steps. Model inversion attacks use AI model output to reverse engineer and reconstruct input data. In the context of M&A transactions, this could result in the exposure of confidential data or sensitive financial information, presenting elevated risks in the transaction.

Adversarial Attacks

An adversarial attack involves making small, often imperceptible changes to the input data to mislead AI models and generate incorrect outputs. This can cause unnoticed financial inconsistencies, disregarded liabilities, or misinterpreted asset values, which can substantially impact the final deal.

Transfer Learning Vulnerabilities

The use of pre-trained AI models for new related tasks is known as transfer learning. Although this method can be time-saving and cost-efficient, it can also introduce vulnerabilities. Pretrained models may have embedded biases or flaws, posing unforeseen risks and challenges that appear after an acquisition or merger.

Trojan Attacks

In certain cases, attackers can introduce hidden malicious behaviors into an AI model, which can only be activated under specific conditions. In mergers and acquisitions, this could result in an AI system that appears to function properly during the transaction but programmatically switching to a mode that produces misleading output after the acquisition, resulting in financial losses or strategic missteps.

Mitigating AI Threats

As businesses increasingly integrate AI into their operations to drive innovation and optimize processes, the realm of possibilities widens. As AI systems become more intricate and their impact on businesses becomes more profound, their risks become increasingly complex. Successfully navigating these risks requires an approach that is adjusted to the distinct challenges presented by AI.

Deep Technical Audits

Although traditional due diligence is thorough, it may not be sufficient for AI systems. A comprehensive evaluation would involve performing technical audits that focus on the intricacies of AI models, data sources, and data pipelines. These audits should encompass an in-depth assessment of data quality and integrity, as well as the robustness and validity of AI models in diverse scenarios. Additionally, it is important to examine algorithms for biases, vulnerabilities, and ethical concerns to identify and mitigate the risks associated with AI. Such audits should:

- Evaluate the quality and integrity of the data that flows into AI systems.
- Assess the robustness and validity of AI models, ensuring that they perform as expected in diverse scenarios.
- Examine the algorithms for biases, vulnerabilities, or ethical concerns.
- Assess the entire data pipeline to ensure its integrity.

AI Security Specialists

Given the intricacies and complexities of AI threats, general cybersecurity knowledge may not suffice in this rapidly evolving and specialized AI security field. It may be necessary to engage professionals who specialize in AI security to assess and address vulnerabilities and ethical concerns. Key areas that require specialized AI security expertise are as follows:

- Insights into the latest threats and vulnerabilities specific to AI.
- Recommendations on best practices and mitigation strategies tailored to AI systems.
- Expert evaluation of AI models ensures that they adhere to industry standards and best practices.

Continuous Monitoring

Due to the dynamic nature of AI systems, continuous monitoring is necessary to ensure optimal performance and security. This ensures:

- Early detection of vulnerabilities or anomalies in AI system behaviors.
- Timely updates and patches to AI models.
- Monitoring of data sources and ensuring that the quality and integrity of data for training AI systems remain uncompromised.

Training and Awareness

It is essential to recognize that, while technology can assist in protecting AI systems, humans are irreplaceable. A knowledgeable team serves as an initial defense against threats specific to AI. Companies should focus on the following:

- Educate teams about the unique challenges and threats posed by AI.
- Provide guidelines on best practices when working with AI systems.

- Foster a culture of continuous learning so that teams are up to date with the latest advances in AI security.

Cybersecurity and AI in M&A

As businesses increasingly rely on AI systems, their security must be incorporated into cybersecurity plans. In the context of mergers and acquisitions, cybersecurity due diligence must evolve to encompass the assessment of AI assets, including their value and vulnerabilities, in addition to traditional IT assets. This includes:

- Evaluating the security protocols for AI data storage and processing.
- Assuring the security of the data processing pipeline that flows into the AI models.
- Assessing the risks associated with proprietary AI algorithms.
- Understanding the long-term implications of integrating AI systems post-acquisition.

Conclusion

The complex nature of M&A deals means that cybersecurity concerns and risks cannot be avoided. As we have seen, cybersecurity breaches have led to loss of value and weakened the strategic outcomes for many M&A transactions. The Mergers & Acquisitions Cybersecurity Framework (MAC-F) solves this problem by taking a comprehensive approach to identifying risks and security deficiencies throughout the deal's lifecycle. It provides related financial transparency, quantifying exposures to maximize deal value by proactively reducing risk.

A key advantage of MAC-F is its adaptability. As artificial intelligence and other emerging technologies are adopted by companies and in M&A activities, the framework is designed to adjust and continue to mitigate new threats. Its modular design ensures it can evolve alongside industry. By taking a holistic view of cybersecurity and enabling informed risk management, the MAC-F allows organizations to execute M&A strategies with confidence despite the complex and ever-changing security landscape.

Visit *https://www.mac-framework.com for more information on using the MAC-F for your organization.*

Glossary

Acquisition - The purchase of one company by another.

Advanced persistent threat (APT) - A long-term, targeted cyberattack to gain unauthorized network access.

Adversarial attacks - Manipulating input data to mislead AI systems.

Application security - Protecting software applications from threats and vulnerabilities.

Artifacts - Documentation or outputs that provide evidence of security measures.

Artificial intelligence (AI) - Computer systems that exhibit human-like intelligence.

Asset management - Tracking, managing, and protecting an organization's digital and physical assets.

Asymmetric risk - A situation where minor vulnerabilities can have major consequences.

Audit - A systematic examination of an organization's compliance and security controls.

Availability - Ensuring systems and data are accessible and functioning for authorized users.

BC/DR - Business continuity and disaster recovery planning.

Biometrics - Using human characteristics for authentication, such as fingerprints.

Breach - A security incident where data or systems are compromised.

Buyer - The company or entity acquiring or purchasing another company.

Capital costs - One-time expenses required to implement security measures.

CCPA - California Consumer Privacy Act, a State of California privacy law.

Chief Information Security Officer (CISO) - An executive responsible for overseeing enterprise cybersecurity strategy and operations.

Closing - The final stage of an M&A deal where legal transfer of assets occurs.

Compliance - Conforming with regulations, mandates, and legal obligations.

Confidentiality - Ensuring information is only accessible to authorized parties.

Corporate governance - The system of policies, processes, and people directing and controlling an organization.

Cyberattack - A malicious attempt to damage, disrupt, or gain unauthorized access to computer systems.

Cybersecurity - Protecting systems, infrastructure, and data from digital threats.

Cyberthreat - A person, action or activity that may damage or compromise systems and data.

Data encryption - Encoding data to protect its confidentiality and integrity.

Data loss prevention (DLP) - Tools and policies for preventing unauthorized data exfiltration.

Data migration - Moving data from one location or system to another.

Data poisoning - Corrupting AI training data to skew outputs.

Deal impact - The effect of cyber risk on the valuation or viability of an M&A deal.

Definitive agreement - The final binding contract outlining the terms of an M&A deal.

Denial of service (DoS) - Flooding systems to make them inaccessible to users.

Detect - Monitoring for threats and vulnerabilities in the environment.

Distributed denial of service (DDoS) - Using multiple sources to flood and overwhelm a system.

Due diligence - Investigating and evaluating an investment or acquisition target.

Endpoint - Devices like laptops and mobile phones that connect to a network.

Endpoint detection and response (EDR) - Tools to monitor endpoints and remediate threats.

Escalation of privilege - Abuse of authorized access to gain greater privileges.

Financial audit - Examining an organization's financial statements and accounting practices.

Firewall - A network security device that controls and filters traffic.

GDPR - General Data Protection Regulation, an EU data privacy law.

Identity management - Managing access to systems by verifying user identities.

Incident response - Detecting, responding to, and recovering from security incidents.

Information security - See cybersecurity.

Insider threat - A security risk from current or former employees and contractors.

Intellectual property - Creative and proprietary information, such as trade secrets or patented designs.

Internet of Things (IoT) - Networked devices and sensors that connect to the Internet.

Letter of intent (LOI) - A document outlining the intent to enter into a business transaction.

Liabilities - Legal responsibilities, debts, or obligations.

Log management - Collecting, storing, and analyzing log data.

MAC-F - Mergers and Acquisitions Cybersecurity Framework.

Malware - Software intended to infiltrate and compromise computer systems.

Management focus - Leadership oversight and governance of cybersecurity.

Merger - The combination and integration of two companies.

Model inversion attack - Reconstructing AI training data from model outputs.

Monitoring - Tracking systems and network activity for anomalies.

Multifactor authentication - Requiring two or more credentials to verify a user's identity.

Network security - Protecting the systems and data accessible through networks.

Non-disclosure agreement (NDA) - A contract requiring parties to keep shared information confidential.

Organizational security awareness - Educating employees about cybersecurity and threats.

PCI DSS - Payment Card Industry Data Security Standard.

Penetration testing - Simulated attacks to identify vulnerabilities.

Permissions - The level of access granted to users or systems.

Phishing - Fraudulent emails or sites that steal user data.

Physical security - Safeguards like locks and alarms to protect facilities.

Policies - Documented cybersecurity standards and expected behaviors.

Post impact - The effect of a cyber risk after an M&A deal is completed.

Prevent - Cybersecurity controls that stop threats before they cause harm.

Privileged access - Specialized access granted only to certain authorized users.

Procedures - Documented instructions for executing cybersecurity processes.

Ransomware - Malware that encrypts data until a ransom is paid.

Recurring costs - Ongoing annual expenses to maintain security measures.

Red team - Cybersecurity personnel who emulate attackers.

Respond - Capability to contain, eradicate, and recover from incidents.

Risk - The likelihood of a vulnerability being exploited and its impact.

Risk assessment - Identifying and evaluating risks to prioritize responses.

Rootkit - Software tools that enable administrator-level access.

Security operations center (SOC) - A team that monitors and responds to security alerts.

Seller - The company or entity that is being acquired or merged with another entity.

SIEM - Security Information and Event Management system.

Social engineering - Manipulating people to divulge confidential information.

Spear phishing - Highly targeted phishing attacks.

Stakeholders - Individuals and entities who have an interest in an organization.

Strategic roadmap - A plan outlining steps to achieve strategic goals.

Systems security - Safeguarding infrastructure, hardware, and software from threats.

Target - The company being acquired in an M&A deal.

Threat - A person, action or activity that may damage or compromise systems and data.

Threat actor - An individual or group operating with malicious intent.

Threat model - A representation of threats and related countermeasures.

Transfer learning - Applying AI models trained for one task to a related task.

Trojan horse - Malware disguised as legitimate software.

User awareness - See organizational security awareness.

Valuation - The assessed worth of a company.

Vendor risk - Cyber risks introduced by third-party partners.

Vulnerability - A weakness that can be exploited to compromise security.

Vulnerability management - Identifying and remediating security weaknesses.

Zero-day exploit - Attacks that leverage unknown or unpatched vulnerabilities.

Made in the USA
Las Vegas, NV
14 April 2026

45357557R00089